Designing for Minimal Maintenance Expense

QUALITY AND RELIABILITY

A Series Edited by

Edward. G. Schilling
Center for Quality and Applied Statistics
Rochester Institute of Technology
Rochester, New York

1. Designing for Minimal Maintenance Expense: The Practical Application of Reliability and Maintainability, *Marvin A. Moss*

In preparation

2. Quality Control for Profit, Second Edition, Revised and Expanded, *Ronald H. Lester, Norbert L. Enrick, and Harry E. Mottley, Jr.*

Designing for Minimal Maintenance Expense

The Practical Application of Reliability and Maintainability

Marvin A. Moss
Syscon Corporation
Montrose, California

MARCEL DEKKER, INC.
New York and Basel

ISBN 0-8247-7314-4

MARCEL DEKKER, INC.
270 Madison Avenue, New York, New York 10016

Current printing (last digit):
10 9 8 7 6 5 4 3 2 1

PRINTED IN THE UNITED STATES OF AMERICA

To Luci—in appreciation

Introduction to the Series

The genesis of modern methods of quality and reliability will be found in a simple memo dated May 16, 1924, in which Walter A. Shewhart proposed the control chart for the analysis of inspection data. This led to a broadening of the concept of inspection from emphasis on detection and correction of defective material to control of quality through analysis and prevention of quality problems. Subsequent concern for product performance in the hands of the user stimulated development of the system and techniques of reliability. Emphasis on the consumer as the ultimate judge of quality serves as the catalyst to bring about the integration of the methodology of quality with that of reliability. Thus, the innovations that came out of the control chart spawned a philosophy of control of quality and reliability that has come to include not only the methodology of the statistical sciences and engineering, but also the use of appropriate management methods together with various motivational procedures in a concerted effort dedicated to quality improvement.

This series is intended to provide a vehicle to foster interaction of the elements of the modern approach to quality including statistical applications, quality and reliability engineering, management, and motivational aspects. It is a forum in which the subject matter of these various areas can be brought together to allow for effective integration of appropriate techniques. This will promote the true benefit of each, which can only be achieved through their interaction. In this sense, the whole of quality and reliability is greater than the sum of its parts, as each element augments the others.

The contributors to this series have been encouraged to discuss fundamental concepts as well as methodology, technology, and procedures at the leading edge of the discipline. Thus, new concepts are placed in proper perspective in these evolving disciplines. The series is intended for those in manufacturing, engineering, marketing and management as well as the consuming public, all of whom have an interest and a stake in the improvement and maintenance of quality and reliability in the products and services that are the lifeblood of the economic system.

The modern approach to quality and reliability concerns excellence. Excellence when the product is designed, excellence when the product is made, excellence as the product is used, and excellence throughout its lifetime. But excellence does not result without effort, and products and services of superior quality and reliability require an appropriate combination of statistical, engineering, management and motivational effort. This effort can only be directed for maximum benefit in light of timely knowledge of approaches and methods that have been developed and are available in these areas of expertise. Within the covers of this series, the reader will find the means to create, control, correct and improve quality and reliability in ways that are cost effective, that enhance productivity, and that create a motivational atmosphere that is harmonious and constructive. It is dedicated to that end and to the readers whose study of quality and reliability will lead to greater understanding of their products, their processes, their workplace, and themselves.

Edward G. Schilling

Preface

This book was written to help designers of commercial and industrial hardware items to improve the marketability of their products by reducing the user's ultimate cost of ownership. Maintenance expenses totaled over the useful lifetime of a product will in many instances exceed the original purchase price. For some commonly used products, the ratio of maintenance expense to acquisition cost may reach two to one, three to one, or even higher. If this statement seems an exaggeration, just check the maintenance records for your automobile. The motorist who buys a new car and drives it for ten years can expect to pay out between one and two times the sticker price just for upkeep and repairs.

Savings in product maintenance costs can be translated into customer goodwill and reduced warranty expense. Lifetime maintenance costs are determined by four things: the different ways in which the product can fail to perform properly, the frequencies with which these different types of failures occur, the nature of the repairs required for correcting each type of failure, and the extent of the routine servicing prescribed by the manufacturer. The design analysis and maintenance planning procedures described in Part II of this book will enable you to identify potential failures, minimize their occurrence, and make maintenance simpler and easier to perform. These procedures can be applied profitably to items as simple as a power lawnmower or to systems as large and complex as a global telecommunications network. The procedures will also generate the information needed for preparing repair and overhaul manuals, and for planning factory-supported service after sales.

Part III is intended to serve as a handy reference for reliability and maintainability analysts and logistics planners, as well as a guide for designers. Here they will find methods for modeling complex systems and predicting their RMA (reliability/maintainability/availability) attributes.

The discussions in Part I of reliability, maintainability, and availability provide the background for answering questions that inevitably come up when applying the procedures in Parts II and III. If you like, you can skim the discussions and move right on to the procedures, coming back for explanations whenever necessary. These explanations and procedures have been distilled from sophisticated techniques evolved during three decades of developing advanced-technology systems for military and spacecraft applications. Because the book has been written primarily to satisfy the needs of the pragmatic designer, mathematical derivations have been kept to an absolute minimum. The few equations presented employ no math more complicated than familiar algebraic equations, sprinkled with an occasional exponential or summation expression.

Many designers seem to think that R&M (reliability and maintainability) are black arts practiced by high-level specialists—that these mysterious disciplines are completely beyond the designers' grasp and outside the scope of their work. Nothing could be further from the truth. The fundamentals of R&M are based on common sense, and the designer can verify their validity from experience with the ways in which hardware behaves. In a sense, R&M merely structures something the designer already knows intuitively. But an understanding of R&M principles enables the designer to apply them more thoroughly and efficiently. Of course there are complex design problems that call for the attention of R&M specialists, but even here it is the designer who must implement the recommendations of the specialist, and designers can't do this effectively if they don't speak the specialist's language.

Marvin A. Moss

Contents

Designing for Minimal Maintenance Expense

I

Fundamentals: Reliability, Maintainability, and Availability

Reliability and maintainability are a pair of closely related disciplines that have become standarized engineering functions in the high-technology industries. In many technical organizations the two disciplines are grouped together because they deal with the same design elements and employ the same mathematical basis in their approach—furthermore, data generated in reliability analyses serves as inputs to maintainability analyses. No concensus has been reached on a name for the combined reliability and maintainability (R&M) function, but design assurance seems appropriate because the joint objective of R&M is to assure the end-user that the performance designed into the product—the performance he paid for—is available whenever he wants it.

As one R&M engineer remarked, his job parallels that of a QA (quality assurance) engineer in that each of them best serves his employer by acting as the customer's representative within his own area of responsibility, because customer satisfaction with the product is essential to the continued prosperity of the company. On the organization chart of the ideal high-technology manufacturing concern, the managers of the two functions, QA and R&M, would be shown as reporting to a director of "Product Assurance."

Achieving and maintaining a satisfactory level of product quality is predicated on the assumption that the design is adequate for the application; therefore, the responsibility of the quality assurance organization is to see to it that production units adhere to the design. The term, quality, as we use it in industry, is a product characteristic built into the item by the way it is manufactured. Reliability and maintainability, on the other hand, are characteristics entirely inherent in the product's design and are, therefore, the responsibility of the company's engineering department.

Undertaking the development of an advanced-technology product is invariably risky because it entails the pioneering of unknown territory. Advanced-technology industries are, therefore, very much concerned with risk management, which is the anticipation and control of risk factors. The reliability and maintainability attributes of a proposed design are critical elements of the risk that will be assumed if it is decided to translate that design from paper into hardware.

Although the magnitude of the risk presented by a state-of-the-art design is considerably less than that of its high-technology counterpart, the extent of the risk that can be accepted in designing for the marketplace is correspondingly less, because the customer is unwilling to share any part of it. The Army may be willing to underwrite the risk of producing a new kind of self-guided antitank missile, but the householder does not expect to pay extra for a power lawnmower that mows 17% faster than last year's model. After all, he can already buy one almost as fast from your competitor.

As you probably suspected, designing for minimal maintenance costs falls under the heading of maintainability, rather than under reliability. However, to fully grasp and properly apply the principles of the former, you must first learn the principles of the latter. Because the user of a product needs a simple and convenient means of expressing the combined effect of product reliability and maintainability in terms directly related to the way in which he uses that product, a design attribute called availability has been formulated. These three attributes—reliability, maintainability, and availability—(RMA for short) are best understood when considered in that order, and each has been given a chapter to itself in Part I of this book.

The extent to which a designer must familiarize himself with the material in Part I in order to utilize the procedures of Part II depends upon the complexity of the product he is analyzing. A fundamental understanding of the following should be adequate for dealing with relatively simple products, a category that includes the bulk of all products in terms of numbers of units in common use: the concept definitions in Sections 1.3 (Chapter 1) and 2.1 (Chapter 2), and the explanations of fundamental equations in Sections 1.4, 2.3, and 3.1. Mastery of these will get you started thinking like a RMA analyst. But be warned—long familiarity with the product is not an assurance of its simplicity. Although no consumer product is more familiar to most of us than the family automobile, the auto constitutes a design of utmost complexity to the reliability/maintainability specialist. That's why we use it as our principal source of examples throughout Parts I and II.

1
Reliability

1.1 ORIGINS OF THE RELIABILITY CONCEPT

Reliability is one of those fundamental ideas that everyone seems to grasp intuitively. Why then do technically-minded people find it so very difficult to define reliability in precise terms? The confusion stems from differences in the kinds of situations to which the reliability concept is applied. To take a familiar example, what do you mean when you say that your automobile is reliable? Do you mean that it always starts right off in the morning when you want to leave for work? But, what if the battery has become too weak to turn the engine over because you've neglected to maintain its electrolytic fluid at the proper level? Could this starting problem be considered as evidence that the dealer sold you an unreliable machine?

No? Well, then, perhaps you mean that your auto always gets you where you want to go, once you manage to get it started. But, suppose that your daily trips have been interrupted several times recently, once by a broken fanbelt, another time by a leaking radiator hose, and just yesterday by a nearly flat tire. Is this automobile unreliable? Of course not—again the fault was yours for failing to check these items regularly, as recommended in the owner's manual resting in the glove compartment.

Now, let's make the question a bit more subtle. Suppose that you've serviced the auto just as the manufacturer recommended, but you've had three breakdowns in the past three weeks, nevertheless. First, the water pump seized. Then, the driveshaft universal joint broke. Finally the transmission began slipping badly. Do we have an unreliable product here? Perhaps, but we can't overlook the fact that this is a car with more than 80,000 miles on the odometer, and these are the first major problems you've experienced with it in the past three years.

Perhaps the best introduction to the meaning of reliability can be found in a couple stanzas of an old poem by Oliver Wendell Holmes, Sr., called "The Deacon's Masterpiece, or the Wonderful One-Hoss-Shay."

Now in building chaises, I tell you what,
There is always somewhere a weakest spot,--
In hub, tire, felloe, in spring or thill,
In panel, or crossbar, or floor, or sill,
In screw, bolt, thoroughbrace,--lurking still,
Find it somewhere you must and will.--
Above or below, or within or without,--
And that's the reason, beyound a doubt,
A chaise breaks down but doesn't wear out.

But the Deacon swore (as Deacons do,
With an "I dew vum," or an "I tell yeou,")
He would build one shay to beat the taown
'n' the keounty 'n' all the kentry raoun',
It should be so built that it couldn' break daown,
--"Fur," said the Deacon, "'t's mighty plain
Thut the weakes' place mus' stan' the strain;
'n' the way t' fix it, uz I maintain,
Is only jest
T' make that place uz strong uz the rest."

In the Deacon's view, a reliable machine was one that wears out instead of breaking down, and the way to avoid breakdowns is to make "...that place uz strong uz the rest." The Deacon probably didn't realize it, but he had discovered a fundamental principle in designing products for high reliability. We'll dissect that fundamental principle a bit further on. For now, we can proceed from the Deacon's ground rule to formulate our first definition of *reliability* as follows:

> Any reliable device, whether it be a horse-drawn chaise or any other kind of mechanism, is one that continues to perform its intended function throughout its intended useful lifetime, regardless of adverse operating conditions.

But, it must be hastily added that a product cannot be expected to function reliably if it is not maintained properly or if the adverse operating conditions to which it is subjected exceed reasonable limits.

A particular circumstance made the Deacon's rule a practical guideline for his time; namely, the depth of the experience already possessed by coach builders and blacksmiths. Through a trial-and-error process extending back several centuries, a vast body of knowledge had been accumulated on methods for dealing with the structural stresses that must be withstood by the various components of horse-drawn vehicles. These were, principally, methods for selecting the materials and configurations best suited for fabricating parts that would withstand those stresses. (Of course, today's designers would feel that the Deacon's masterpiece must have been greatly overdesigned to last for a century without a breakdown; ten to fifteen years is generally considered to be adequate and far more cost-effective.)

With the advent of the machine age at the beginning of the nineteenth century, design techniques became somewhat formalized, in order to accommodate the mushrooming development of products employing increasingly complex mechanisms. A feature of this new technology was a trend toward standardizing commonly used parts, such as fasteners and bearings. As standard parts types gained general acceptance, the users and manufacturers of those items noticed that units of a given type tend to break or wear out in the same way, and that the duration of the part's useful life (i.e., operating time before failure) can be directly correlated to the severity of the stresses imposed on the part by the application to which it is put.

The correlation between application loading and useful operating life before wearout is particularly noticeable with bearings. Since a bearing failure generally results in the failure of the product in which it is installed, a great deal of effort has been expended by bearing manufacturers on identifying the factors that induce failure. Eventually it was found that the relationships between these factors could be expressed in the form of mathematical equations that could be passed along to the designer to guide him in selecting the proper type and size of bearing for a given application. A typical equation for estimating ball bearing life shows the rated life to be inversely proportional to the rotational speed of the inner ring and the cube (third power) of the applied radial load. The rotational speed and radial load determine the dynamic mechanical stress factors acting on the bearing.

From extensive experience arose the generalization that the useful life of any part is inversely proportional to the magnitudes of the stresses imposed upon it. In the general run of product applications, the stresses to which various components of a product are subjected tend to be of roughly the same magnitude, when considered over the life span of the product. From these premises it can be argued that the reliability of a product is no better than the reliability of its least reliable component part, since the product will fail whenever any one of its components fails. The deacon and coach builders before him had been aware of this. What they lacked was a method for predicting a product's reliability in quantitative terms and, thereby, determining the most cost-effective way of improving its reliability. When that method had been found, it became the core of the reliability discipline.

1.2 RELIABILITY BECOMES AN ENGINEERING DISCIPLINE

Considering the rapid evolution, over the past century, of analytical prediction techniques such as structural stress analysis, and the development of methods for applying the knowledge thus gained to the optimization of product designs, it seems almost inevitable that a body of theory would evolve for predicting and optimizing product reliability. Such a body of theory had, indeed, been synthesized by the end of the Second World War. But formulation of the theory did not immediately stimulate the development of a practical methodology for its application. The idea of reliability as a formal design consideration did not enter the working vocabulary of product development engineers and designers.

The great breakthrough in applying reliability theory came in the latter 1950s, in the form of specific requirements stipulated by military contracts for major missile weapon systems. Within a few years, reliability requirements became a familiar part of procurement specifications issued by the military services for new hardware. At first, these requirements were stated in terms of the probability of successfully completing a prescribed mission. The reliability design objective was expressed as a non-dimensional value, for probability of success, that can approach but never reach unity (1.0), or as a percentage value that can approach but never reach 100 percent.

This probability-of-success approach had originated in the aviation industry between the First and Second World Wars, when great interest existed in measuring the reliability advantage offered by the use of multiple-engine designs over single-engine arrangements. But these studies did not result in a methodology that could be applied to product design. Credit for developing a reliability concept upon which quantitative analysis techniques could be based must go to two men, Robert Lusser and Erich Pieruschka, who were attempting to deal with a consistent string of failures haunting the testing of the German V1 missile at Peenemunde during World War II. Every missile test failure had been traced to the failure of a single component, but not necessarily the same component. In keeping with the idea that a chain can be no stronger than its weakest link, corrective efforts had been concentrated on improving the most failure-prone components. But further testing showed that even good components sometimes failed. Thus it could be seen that the reliability of the missile system depended on the reliabilities of *all* its components—a reliability chain is actually *weaker* than its weakest link.

In following up the idea that the reliabilities of all components must be considered when predicting the reliability of a product treated as a system, it had been intuitively assumed that system reliability can be equated to the average of the reliability values of its components. However, accumulating test data indicated that system reliability is actually substantially less than this average. A sudden insight produced the surprising hypothesis that system reliability is not the average but, rather, the product of its component reliabilities. The entire body of reliability theory utilized today derives from that simple statement. Since the reliability value of each component must, by definition, be less than unity, the reliability value of the system must not only be less than unity but, also, less than the value of its weakest component. The discovery stimulated efforts at component improvement to the extent that the reliability of the V1 missile grew from zero to 75 percent.

Calculating the reliability of a system is synonomous with predicting the system's future reliability performance. Because reliability predictions are based on probability theory, these predictions are valid only for statistically significant populations of like items—the larger the population, the more accurate the prediction. Consequently, the analyst may feel very confident in predicting that 90% of a production lot of missiles will complete their missions successfully, but a 90% prediction for the success of a single manned-spacecraft mission has no physical meaning that can be tested by comparison with the actual outcome—the outcome can be measured only in terms of 100% or 0% success.

Nevertheless, a prediction is needed for the single-mission case, for two reasons. First, by providing a quantitative estimate of the likelihood of success, the prediction supports the decision-making process as to whether the mission should even be attempted. Second, the analysis on which the prediction is based serves to identify opportunities for improving the chances for success by making appropriate design changes. Nonetheless, the fact remains that the quantitative reliability value predicted for a single mission has meaning only in a psychological sense, analogous to the confidence felt by a gambler in predicting the outcome of a single roll of the dice.

1.3 FUNDAMENTAL RELIABILITY DEFINITIONS

At this point, it is necessary to introduce certain fundamental reliability concepts and terms used throughout this book. These concepts are presented as a series of logically-sequenced definitions that agree essentially, with general usage but are formulated more strickly, in order to order to avoid possible misinterpretation. It is recommended that you read these definitions thoughtfully—this is not the section of Part I to be skimmed.

1.3.1 System

A *system* is any composite of hardware items that work together to perform a designated mission or set of related missions. (It must be pointed out that some of the most essential components of information processing and transfer systems are "software," not "hardware," but the treatment of software reliability falls outside the scope of this book.)

A complex system may be made up of two or more groupings of hardware items, each of which accomplishes a distinct one of the system missions. However, the definition of the mission to be performed in a particular case is necessarily subject to individual interpretation. Consider, for example, a residential hot-air furnace and air conditioning system. Until recently, heating and cooling were thought of as separate missions because they were performed by completely independent and physically separate systems. Integration of the two kinds of equipment has been found advantageous for several reasons, one of which is that the same air circulation ducting may be used for both missions. But this raises a question: are there now two missions or one?

It could be said that the integrated system has only one mission, that of maintaining a comfortable atmosphere in an inhabited area. If so, it is necessary, for the purpose of evaluating the performance and the reliability of the system, to subdivide the mission into at least three related but distinct sub-missions: (a) control of the air's sensible temperature; (b) control of the air's relative humidity;

(c) distribution of the air thus controlled, and, if the system includes air filtering, (d) control of the air's contamination level. To the extent that each of these submissions is performed by a distinct group of hardware items, that group can be designated as a subsystem. For reliability analysis convenience, each subsystem is treated as a system in its own right.

1.3.2 Circuit

A *circuit* is any grouping of hardware items, within a system or subsystem, that operate together to perform a single distinct function contributing to the accomplishment of the system's mission. (In the context of this book, the term, circuit, is not restricted to the familiar electrical circuit but can also be applied to linkages of mechanical, hydraulic, thermal, or other kinds of hardware.) Components of the circuit include not only the parts whose operations contribute directly to accomplishment of the circuit's function(s) but also the wiring, piping, shafting, tubing, ducting, belting, etc., that interconnect those parts to form the circuit.

In the case of the integrated atmospheric comfort maintenance system (i.e., the aforementioned air conditioning system), two distinct functions—heating and cooling—are performed by components of the air temperature control subsystem. The group of components involved in the heating function can be designated as the heating circuit, and those in the cooling function as the cooling circuit. Some of the components in the air temperature control subsystem may be involved in both functions, and may thus be considered to be utilized by both circuits on a time-sharing basis.

The relationship between "circuit" and "function" is fundamental to the proper implementation of RMA analysis techniques, so let's explore it further with our favorite example. Ask yourself the question, what are the functions of the electrical circuitry of an automobile? As you will probably discover when completing a preventive maintenance analysis (Chapter 4) on one of your own product designs, defining the functions performed by the major circuits can sometimes by quite difficult. You may find yourself changing the function definitions several times in order to make the answers to other questions in the analysis consistent among themselves.

Getting back to the question of automobile electrical circuitry, it appears that the function performed by the circuitry must be ranked in order of relative criticality with regard to the impact of a circuit failure on the success of the mission. The primary function is driving the starter motor—if the circuit cannot provide sufficient power, the automobile cannot be started—the mission is a failure at the outset. The electrical circuiting also provides power to various assemblies required for safety purposes—headlights, tail lights, running lights, horn, windshield wipers, rear window defroster, etc. Note that, although safety is of transcendent importance to the users of the product and, therefore, to the manufacturer, it is only incidental to the mission, which is to transport people and their goods—consequently, powering the safety devices must be classed as secondary functions. The electrical circuitry also performs a tertiary group of functions by providing power to convenience accessories such as interior lights, the radio/cassette player, the cigarette lighter, etc.

To sum up what we've said so far, identifying the mission(s) to be performed by the product is the first step in a logical design approach; the next step is to break down each mission into the functions essential for its accomplishment. When this has been done, those abstract concepts called functions can be transformed into circuits composed of interconnected hardware items. Then the circuits are arranged into subsystems, and those subsystems packaged to comprise the end product. The system thus depicted can now be analyzed to predict its inherent performance capabilities, and the design modified as necessary to optimize those capabilities within the constraints of producibility and manufacturing cost.

One of the designer's principal tasks is to analyze the performance capabilities of a product so that he can devise design modifications that will improve that performance. Performance analysis of a complex product can be greatly simplified by modeling the design as a system made up of functional circuits. The same system/circuits model is also used for analyzing and optimizing the product's reliability, as we will see in Part III.

In simple designs, each circuit component is required to perform only one action. A more sophisticated approach utilizes components that combine capabilities (e.g., a switch-type circuit breaker that enables an operator to turn the circuit on and off manually, in addition to protecting the circuit against current overloads). An extension of the sophisticated approach shares components between circuits (e.g., routing through a common circuit breaker those circuits whose combined normal current drain matches the rated capacity of the circuit breaker; excessive current in any of the circuits will trip the circuit breaker).

For systems that include circuits sharing components, functional relationships will not always be synonomous with the physical-interconnection relationships. Consequently, two different logically-consistent schematic representations may be devised for the system, one based on functional relationships and the other on interconnections, but only the functionally-related representation can be used for reliability analysis.

1.3.3 Part

A *part* (or *piece-part*) is any hardware item that cannot be disassembled into subordinate components without disrupting permanent physical bonds.

1.3.4 Assembly

An *assembly* is any structural or dynamically functional item that can be disassembled into two or more subordinate components without disrupting permanent physical bonds. The components of an assembly may be any combination of subordinate assemblies (subassemblies) and piece-parts.

The term, *structural*, in this context refers to the role played by passive components that support the dynamically functional components directly engaged in the implementation of circuit functions. Various types of structural components serve to: (a) provide support against mechanical stresses, including those produced by the force of gravity; (b) protect against adverse environmental conditions, such as wind, rain, and sunshine; (c) facilitate handling of the packaged circuitry; and (d) maintain precise dimensional positioning of geometrically related components.

1.3.5 Assembly Hierarchy

Assembly hierarchy describes the organization of system hardware elements into assembly packaging levels descending from the top (or system level) on the basis of functional and/or structural interrelationships. (An assembly hierarchy is frequently referred to as an "assembly tree," but is depicted on charts as appearing like the *roots* of a tree, each root terminating in a piece-part.)

Reliability analyses are performed in reverse hierarchic sequence: first, the reliabilities of the lowest-level (piece-part) components are determined. From these individual piece-part reliabilities, the reliabilities of the next-higher-level assemblies comprised by these piece-parts are calculated, and so on upward until the uppermost level, the system level, has been accounted for.

1.3.6 Malfunction

A *malfunction* is any occurrence, other than intentional shutdown, wherein a system ceases to carry out its specified mission or a circuit ceases to perform its function. A malfunction can be a complete cessation of operation (usually classified as a catastrophic malfunction), a gradual uncontrollable drift of a critical performance parameter beyond specified limits (degraded performance malfunction), or intermittent transient excursions of a critical performance parameter beyound specified limits (may be considered catastrophic).

A malfunction can result from any one of the following circumstances:

1. Faulty or maladjusted component.
2. Functional misalignment or faulty interconnection between interfacing components. (A *functional misalignment* is a circuit malfunction that occurs even though every one of the components of that circuit is operating within the performance limits specified for it. This type of malfunction is attributed to an unfavorable build-up of component performance tolerances, and must be regarded as a defect in the design of the circuit.)
3. Physical damage from an external cause.
4. An erroneous external control signal or, for a transmission circuit or computer, an erroneous data input signal.
5. Control or data signal interference from an external source (e.g., radiated or conducted electromagnetic noise or pulses).
6. An aberration in externally furnished power.

1.3.7 Failure

A *failure* is any malfunction of a system or circuit that persists or recurs until appropriate corrective action is taken. In the case of a system that accomplishes more than one independent mission, a malfunction impairing the ability of the system to accomplish any one of those missions is judged to be a failure, even though the other mission capabilities are not affected.

For reliability analysis purposes, failures are classified as either chargeable or non-chargeable. A *chargeable failure* is a failure that can be attributed to a defect in design or manufacture. Failures attributable to physical degradation normal for the item in its designated application are treated as limitations inherent in the design, and are classified as chargeable. A *non-chargeable failure* is a failure caused by: exposure of the item to an operational, environmental or structural stress beyond limits specified for the design; an error commited by an operator; or improper handling or maintenance. Chargeable failures are so named because the manufacturer is charged with responsibility for those failures and is, therfore, presumed liable for their correction, at least during the warranty period. Only chargeable failures are considered in reliability analyses because the designer has no control over non-chargeable failures.

1.3.8 Failure Source

A *failure source* is the next-lower-level component to which the failure of an item can be isolated. The ultimate failure source is the faulty piece-part or faulty interconnection between piece-parts or assemblies whose physical deterioration or destruction caused the failure.

1.3.9 Failure Mode

A *failure mode* is any one of the failure scenarios that a given system or circuit may exhibit, in terms of the mission(s) or function(s) it performs. Physical deterioration of a piece-part within the generator of an auto electrical circuit can cause the generator to fail in any one of the following modes: (a) severe reduction of output voltage (a degraded performance mode); (b) complete cessation of output (catastrophic mode); (c) complete cessation of output, plus short-circuiting of the battery, thereby destroying it and, possibly, other components of the electrical circuitry (worst-case catastrophic mode).

At the next higher assembly level, that of the automotive system, the corresponding failure modes are more complex in terms of the effect of the mission, because the effects also depend on where, in the automobile's START-RUN-STOP mission sequence, the generator failed. If, as is quite likely, the generator failed in the ordinary catastrophic mode late in the mission sequence, the battery may still retain sufficient charge for a subsequent restart or, at least, operation of emergency lights. Failure in the degraded performance mode can result in somewhat the same consequences, even if it occurs earlier in the mission sequence. But failure in the worst-case catastrophic mode destroys all reserve power.

The differences between failure modes, in terms of corrective maintenance, are scaled in similar fashion. For all three modes, the generator will have to be repaired or replaced. But the extent to which the battery will have to be recharged depends on whether the failure was degradative or catastrophic, and on where in the mission sequence the failure occurred. For the worst-case catastrophic mode, the battery cannot be recharged and will have to be replaced. Other components of the electrical circuit may have to be replaced, as well.

To clarify the relationship between failure mode and failure source, consider an automobile that has suffered a ruptured radiator hose. The failure source is the radiator hose, and the *immediate* failure mode is the structural disruption of that hose under steam pressure because of material deterioration; symptoms include steam emerging from under the hood and a puddle of coolant on the ground beneath the radiator. But the most significant element of this failure mode is the loss of coolant from the engine cooling circuit which, if not detected in time, can result in loss of power and, most probably, substantial damage to the engine. Thus the immediate failure mode can result in further damage if not corrected promptly. Failures of other components of the engine cooling system might be characterized by the same symptoms and the same effect on circuit operation, the only feature unique to each being the immediate failure mode; i.e., the identity of the failure source and the manner in which it failed.

1.3.10 Fault

Fault refers to the physical mechanism of a failure source; usually mechanical wear, chemical corrosion, material aging, or structural deformation or rupture. In the example of an immediate failure mode given in Paragraph 1.3.9 above, the physical mechanism was the structural disruption of the radiator hose.

1.3.11 Service Life

Service life designates the time span during which a product can be expected to operate safely and meet specified performance standards, when maintained in accordance with the manufacturer's instructions and not subjected to environmental or operational stresses beyond specified limits. The service life rating for a given product should represent a prediction that no less than a predetermined proportion of production units will operate successfully for the stated period (in probability terms, a lower confidence limit). A product's service life rating may be tied to a warrantee, in which case the determination of the proportion must take into account the cost of repairing units that fall short of the rating.

Service life may or may not be terminated by a failure occurrence, depending on the kind of product involved and the way it is used. (Service life may be expressed in operating hours, calendar years, duty cycles, or other unit appropriate for the product in its designated application.) The term, operate, in this context refers to performance as subject to duty-cycle limitations and other constraints on operation established by the manufacturer.

From the user's viewpoint, cost-effectiveness must be taken into consideration in determining the effective service life of a product. Consequently, the user may employ a shorter span than the rated service life stated by the manufacturer, on the basis that the product has become more expensive to maintain than to replace.

Certain types of product components are characterized by consistent, sharply-defined calendar or operating life limitations. Calendar life limits are sometimes imposed by irreversible chemical reactions leading to such phenomena as rubber "aging," which is actually caused by the material interacting with oxygen or other reactants in the atmosphere. Operating life limits usually occur as the result of some kind of wearout mechanism. Perhaps the most familiar example of a *limited-life component* is the common incandescent light bulb, whose life can be predicted very accurately. In those cases where a component's expected life is substantially less than that of the product in which it is installed, and the operation of the product depends on the functioning of the component, the product can be expected to fail before its intended service life has been fully expended. However, this failure source can be eliminated by replacing the limited-life component before its predicted life limit has been reached.

1.3.12 Reliability

At last we come to the definition that is at the heart of this chapter—*reliability* is the probability that a product manufactured to a given design will operate throughout a specified period (in some cases the product's related service life) without experiencing a chargeable failure, when maintained in accordance with the manufacturer's instructions and not subjected to the environmental or operational stresses beyond limits stipulated by the manufacturer or set forth in the purchase agreement.

The reliability of a product is an attribute inherent in its design—reliability represents a potential capability that will probably not be fully achieved in actual service unless the product is manufactured exactly as designed, and operated and maintained exactly as prescribed by the manufacturer.

1.4 THE FUNDAMENTAL RELIABILITY PREDICTION EQUATION

Before becoming so deeply involved with definitions, including the definition of reliability itself, we had been exploring the measuring of system reliability in terms of the probability of success of the system completing a designated mission. But, probability of success does not provide a very useful estimate of reliability when evaluating a proposed design. After all, a probability prediction is only an expression of relative confidence in the outcome. A high probability of success may give you a "warm feeling" but doesn't furnish much in the way of design guidance. The work of Dr. Pieruschka on the V1 missile provided a much more practical unit of measurement. His original formulation for evaluating the reliability of an item, whether it be a single part or a system composed of many parts, was subsequently transformed into the expression below.

$$R = e^{-\lambda t}$$

where

R is the probability of the item completing the specified mission successfully;

e is the Napierian (natural logarithmic base;

t is the duration of the mission;

λ is the failure rate of the item throughout the period, t (assumed to be constant)

This simple equation is the foundation of all reliability analysis methodology. The term in the fundamental equation that makes this equation so useful is the *failure rate*, λ (represented by the symbol for the character, lambda, in the Greek alphabet). Every part has a characteristic failure rate pattern which can be mapped from data compiled during laboratory testing or derived from service records. The failure rate pattern characteristic of a given type of part is determined by three factors: (1) the design of the part; (2) the quality standards to which the part is manufactured, and (3) the stresses to which the part is subjected during its service life.

In dealing with circuit failures caused by a given failure source, we are actually looking at physical faults characteristic of the generic type of part involved. The failure rate tends to vary with the intensity of the operational and environmental stresses to which the part is subjected. However, this variation can be manifested in either of two ways. The first way is a proportional relationship that can, in turn, be classified as either linear or exponential. One example of this kind of variation is afforded by the ball bearing, whose failure rate increases linearly with rotational speed or exponentially with radial loading. The *fundamental reliability equation* applies very nicely to either of these proportional variation relationships.

For the other kind of variation, the failure rate remains unchanged at the random failure level despite steadily increasing stress intensity, until the intensity reaches a threshold level, where the failure rate suddenly begins a very steep climb. Typical of this kind of variation is the relationship between structural failure rates and mechanical stresses such as tension, compression, torsion, vibration, and shock. In the simplest instance, if we were to subject to a destructive tensile stress test each one of a group of metal bars taken from the same production lot, we would expect to find that the bars exhibit essentially identical failure patterns—a smooth progressive elongation linearly proportional to the gradually increasing stress intensity until the stress reaches a critical level where the elongation process suddenly accelerates and the bar ruptures. The critical stress intensity—the threshhold level—would be almost identical for all of the bars, with slight variations being accounted for by minor microstructural differences in the metal.

The term, failure rate, is inappropriate in these cases because production units of the part never fail at stress intensities below a clearly evident threshhold level, but virtually always fail at intensities falling within a very narrow band; this band encompasses variations in threshhold level from unit to unit, stemming from manufacturing tolerances. The fundamental reliability equation is valid for cases of the second kind only when the imposed mechanical stresses are kept below threshhold limits so that failure results from wearout, corrosion, or other agent of progressive degradation only.

The failure rate for a circuit or system depends not only on the stresses to which each of its component parts is subjected, but also on the way that those parts are functionally linked together. The simplest case is that of a circuit whose components are linked in such a way that the failure of any one part will cause the circuit to fail; the failure rate for this circuit is the sum of the failure rates for the component parts. This kind of circuit can be depicted schematically by a block diagram in which the blocks representing each component are linked to form a single chain. If one link fails, the entire chain has failed. For reliability analysis purposes, this arrangement is designated as a series-linked model. More complex circuits can be modelled as blocks linked in series-parallel configurations. Methods for modelling and determining the failure rate of series-parallel circuits are presented in Chapter 7.

The reciprocal of the failure rate (i.e., $1/\lambda$) is equivalent to the *mean time between failures (MTBF)*. The reliability inherent in a given product design is most commonly stated as its MTBF, expressed in hours. Failure rates should, therefore, be stated in terms of failures per hour; however, because this would result in writing lots of zeroes between the decimal point and the first significant figure, failure rates are usually written as failures per *million* hours. The MTBF for a given product may be 100 hours, but this does not mean that you can expect the product to fail about every 100 hours—in service, one unit may not fail for 300 hours and then, after repair, fail twice more within the next hour of operation—MTBF is only a *mean* value.

Another point—the MTBF of a product should not be confused with the product's service life rating. A typical MTBF for an automobile might be 100 operating hours (or 4000 miles), while its service life rating is stated to be 2,500 operating hours (or 100,000 miles). The service life for this kind of product is limited by economic factors—with continuing repairs, the auto may run for a million miles but the mounting cost of operation and repair per mile beyond the 100,000-mile mark makes replacement at that point more cost-effective. An incandescent light bulb, on the other hand, can be expected to operate for a stated period, say 1000 hours, and then fail. Its service life is 1000 hours but it cannot be repaired after the first failure.

The MTBF of the light bulb must be derived from a failure rate that represents failures occurring before the rated service life has been expended. Light bulbs fall into the category of "limited-life" items whose MTBF's are based on replacement on or before expiration of rated service life.

1.5 THE FAMOUS "BATHTUB CURVE"

In the previous section (Section 1.4), it was pointed out that the reliability analysis methods generally used today are based on an assumption that the reliability of a hardware item, either a single part or a complex system, can be predicted by means of a simple exponential equation. Implicit in the acceptance of this equation as a mathematical model of reliability behavior is a subordinate assumption to the effect that the failure rate, λ, of the item is constant over the item's entire operating life. But, what if the failure rates of individual functional piece-parts are not constant over the parts' entire useful lives? This would certainly complicate predicting the reliability of a product.

On the basis of lifetime testing conducted on production lots of different kinds of parts, it has been postulated that a curve representing failure rate plotted against hours of operation actually depicts three successive stages of failure phenomena, as shown in Figure 1-1. (This plot, dear to the hearts of reliability engineers everywhere, has come to be known as the *"bathtub curve"* for obvious reasons.)

In the first of the three stages, the failure rate plunges downward rapidly from a very high starting point—this is the "infant mortality" stage. Failures occurring during this stage can be attributed almost entirely to manufacturing defects (i.e., defects in procured materials, in-house processing, or workmanship). Failures caused by manufacturing defects tend to show up almost immediately, accounting for the high starting point of the curve.

As the curve levels off, it enters the second stage, which is a straight segment indicating an essentially constant failure rate whose magnitude is represented by λ in the fundamental reliability equation. In the final stage, the failure rate climbs sharply as units wear out or otherwise deteriorate.

The middle (second) stage, almost invariably the longest of the three stages, is taken as the safe useful service period modelled by the exponential equation, since it enables the manufacturer to avoid the high failure rates of the first (infant-mortality) and third (wearout) stages, claiming the low failure rate of the second stage as the characteristic failure rate for the product.

Once the bathtub curve for a specific part has been derived from test data, it furnishes three very useful pieces of information. The first of these is the duration of the infant-mortality stage; with the duration known, defective parts can be weeded out by actual or

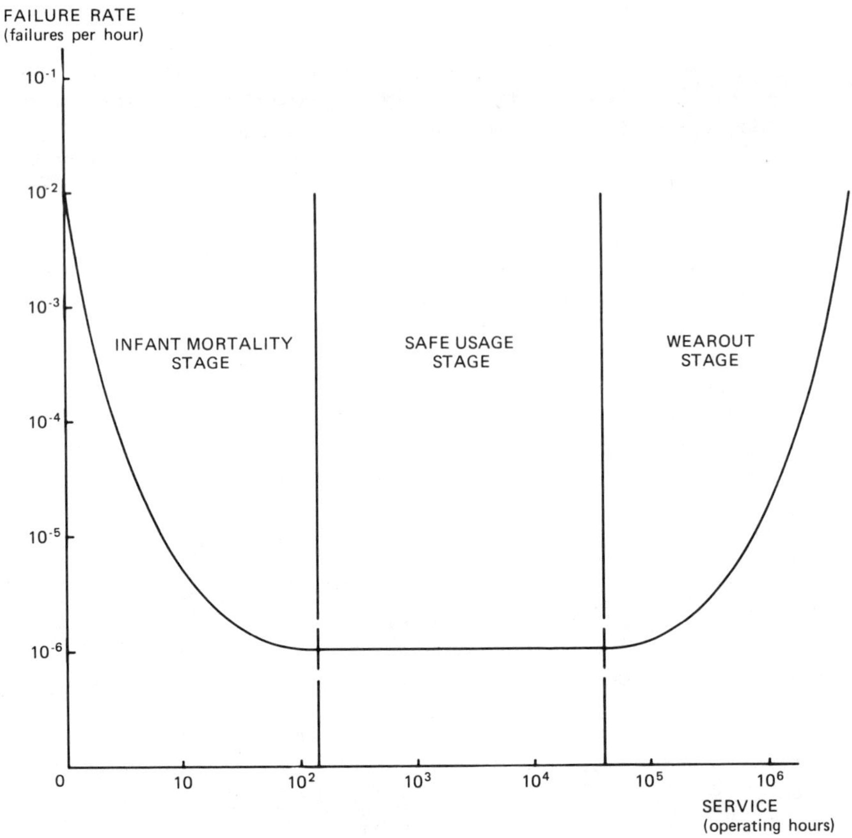

Figure 1.1 Plot of failure rate versus service (the "bathtub curve").

simulated operation of production lots for a period equal to that duration. (This screening procedure is commonly called burn-in). The second bit of info yielded by the bathtub curve is the failure rate prevailing throughout the safe-usage stage; taking the reciprocal of this value (i.e., 1 divided by λ) yields the MTBF (mean time between failures) for the part. The third bit of info is the expected safe useful operating time until the wearout stage begins; the duration of the second stage, minus a safety margin, can be used for scheduling replacement of critical parts to avoid predictable failures. (A critical part is one whose failure will cause the product to fail.) We'll discuss scheduled replacements when we talk about phased maintenance scheduling in Section 6.7.

Unlike the fundamental exponential reliability equation, the bathtub curve can model the reliability characteristic of a generic piece-part type, but not of an assembly, a circuit, or a system. The failure rate predicted for an item comprised of more than one piece-part is a function of the failure rates predicted for its component parts, and is calculated by the methods described in Part III.

The failure rate assigned to each part is taken from the middle segment of its bathtub curve, but with the following points borne in mind:

1. The bathtub curve published for any given part is based on a given set of application conditions, usually a standardized representation of the operating and environmental stresses that the part is expected to sustain during its service life. A change in one of these conditions will displace the curve, relative to the operating time and failure rate axes, and may also significantly alter the shape of the curve or the durations of the three stages represented.
2. The bathtub curve for a given part will probably differ slightly from one production lot to the next, because of minor variations in tooling, workmanship, fabrication processes, materials, and other manufacturing factors. Consequently, the bathtub curve is better depicted as the broad band shown in Figure 1-2, rather than the single line shown in Figure 1-1. The "best-case" boundary represents the optimal combination of manufacturing factors, while the "worst-case" boundary represents the most adverse combination.
3. Completion, by a production lot, of the infant-mortality stage (burn-in) will not necessarily weed out all failures traceable to manufacturing defects. If any remain, they will occur at unpredictable intervals throughout the safe-usage second stage, thereby contributing to the failure rate for that stage.
4. Every failure occurrence must be presumed to have an explainable cause. Those failures occurring during the second stage that cannot be readily identified as manufacturing defects are invariably lumped together as so-called "random" failures because they do not recur frequently enough to justify the cost of identifying the failure mechanism and eliminating it—in the R&M jargon, they do not comprise a definite "failure trend." Indeed, positive identification of certain kinds of failure mechanisms requires elaborate and sophisticated laboratory techniques not always available to

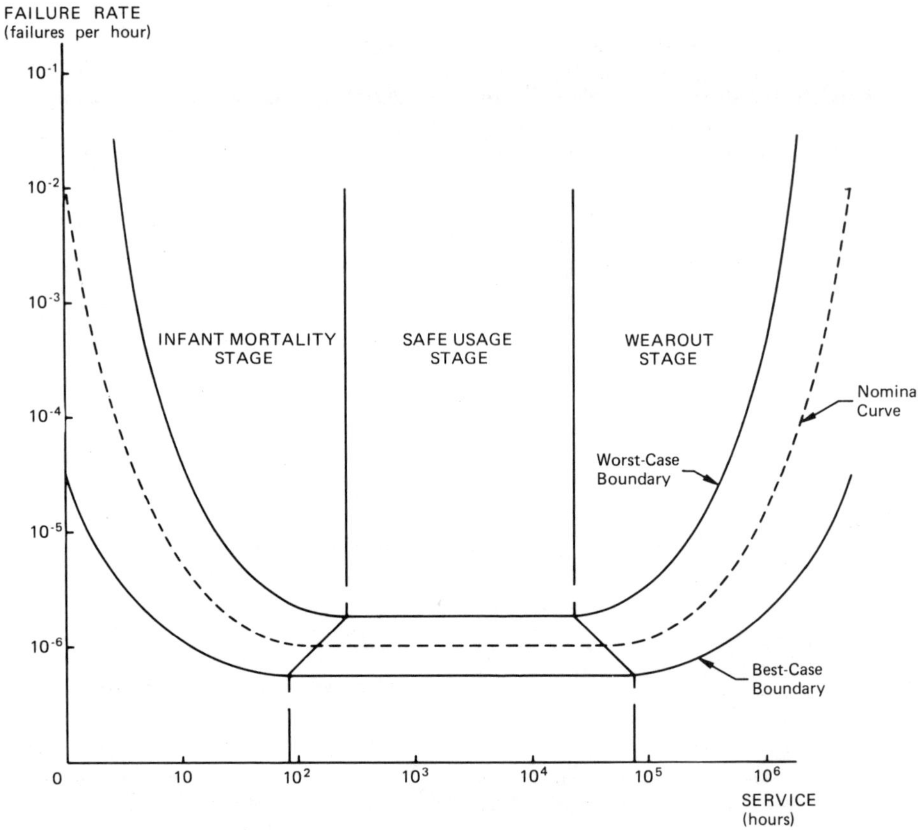

Figure 1.2 Plot of failure rate versus service, reflecting manufacturing tolerances.

a reliability analyst. Moreover, elimination of some kinds of failure mechanisms is not feasible because of economic considerations or state-of-the-art limitations in materials availability or fabrication processes.

5. With the possible exception of solid-state devices, all functional parts appear to possess an inherent degradation mechanism in the form of wearout, corrosion, materials aging, or some other physical phenomenon that will cause the part ultimately to fail in normal service, even if the part is never exposed to stresses beyond the levels for which it was

designed. That's why the bathtub curve always has a third stage.

1.6 THE FIRST GOLDEN AGE OF RELIABILITY

Despite the inception of a practical reliability methodology by the early 1950s, acceptance of reliability design goals as being equivalent in importance to performance goals was resisted by most product development engineers. But when the customer was insistent and willing to pay the extra cost, as the U.S. Government was, reliability became a familiar part of the military designer's working environment.

Pioneering the translation of reliability concepts into working practices took place primarily in the electronics industry. During the postwar era, the electronics industry provided the ideal seedbed for innovative methodologies, for a number of reasons. First, the industry was expanding rapidly because new applications for electronic circuitry were being found almost daily—an expanding industry is always more receptive to new ideas. (Much of this growth had been stimulated by research and development performed for the military services during World War II.) Second, electronic equipment complexity, in terms of the number of components employed in a system, was increasing exponentially, generating horrendous reliability problems. Third, electronic component designs were changing radically because of the need to minaturize equipment for aerospace applications. The miniaturization process was accelerated by the introduction of new component types, particularly solid-state devices, and these rapid design changes introduced additional reliability management problems.

Despite the stepped-up pace in the electronics industry, standardization of electronic components proved easier to implement than standardization of mechanical or other kinds of hardware because relatively few generic component types were needed for building almost any kind of electronic circuitry needed, and because the swift advance of the state of the art quickly obsoleted existing electronic parts inventories.

Military need for improvements in mobile electronic communications, navigation, remote-control, surveillance, and fire-control equipment capabilities offered vast markets for the emerging high-technology electronics industry, but actual demand was severely restrained by previous unfortunate experiences with electronic equipment that had proved embarrassingly unreliable in the field. To break through this failure barrier, the military services underwrote a large-scale effort to incorporate reliability principles into standard design practice. In-house concept studies were undertaken, and research grants were issued to universities and other non-profit institutions, all directed toward the formulation of procedures for implementing the reliability concept. The results of this effort materialized as specific reliability design requirements in hardware development contracts awarded to private industry. Actually, the reliability effort was just one part of a monumental ongoing program and manufacturing processes used for producing items procured by the U.S. Government, particularly the Department of Defense.

1.7 THE CREDIBILITY GAP AND HOW IT WAS BRIDGED

The formative years of the reliability discipline coincided with the genesis of the art of advanced-technology program management for weapon system development. Reliability came to be regarded by the military customer as one of the criteria upon which competing designs proposed for accomplishing a given mission should be judged. Reliability analysis was adopted to serve as one of the methods of controlling development of the selected design. Unfortunately, too much was expected too soon—reliability engineering lost credibility with the customer because some of its practitioners attempted to satisfy demands for accurate predictions before the methodology had sufficiently matured. Consequently, many people in the defense industries came to regard reliability analysis as meaningless "number crunching." But the stigma faded as the gap gradually narrowed between predictions and field results, and as practitioners learned to utilize the predictions as estimates with known limitations, rather than as assured certainties.

The limitations under which reliability analysts labored in those early days were of two kinds. First, the methodology for reliability prediction had not been standardized as to the method used to account for real-world considerations such as effects of severe environmental stresses. Second, component failure-rate data was grossly inadequate because of the lack of mutually-accepted definitions and procedures among the various organizations compiling and publishing the data. Early approaches to data compilation tended to make use of repair information acquired in the course of operating existing equipment. This proved ineffective because the relationships between component reliability and operational and environmental stresses imposed in one design application could not readily be transferred to another application. Furthermore, the delay in making this data generally available also rendered much of it obsolete because of rapid advances in the state of the arts of electronic component and circuit packaging design. Consequently, it is not surprising that independent reliability predictions for a proposed design could vary by wide margins, even though the analysts had all used the same calculation procedures and generally accepted (though different) data sources. (No wonder that skeptics grumbled about meaningless number crunching!)

The impasse was ultimately overcome, for electronic equipment at least, by an approach that combines (a) laboratory testing under strictly controlled conditions, to provide basic component reliability data; with (b) standardized stress factors (multipliers) that can be correlated to most applications. The fruits of this effort appear in Military Handbook MIL-HDBK-217, "Reliability Prediction of Electronic Equipment." Incidentally, Section 4 of the Handbook presents an excellent summation of the kinds of problems encountered in predicting the reliability of electronic equipment. To quote a significant observation made in Section 4, "The ratio of new to tried and true portions of electronic systems is relatively high; therefore, until the new becomes tried and true, its reliability must be suspect."

1.8 LIMITATIONS OF PREDICTION TECHNIQUES

The almost universal acceptance (within prudent limits) of predictions made in accordance with MIL-HDBK-217 serve as a vindication of reliability theory—but those prudent limits must not be overlooked. An examination of Revision D (issued 15 January 1982) outlines those limits, as well as the strengths of the methodology. Finely-shaded distinctions are made between generic component types in the tabulated failure rate data, accounting even for minor differences. All of the application factors that you would expect to influence reliability are also accounted for—operating and environmental stresses, manufacturing quality, and functions performed (if more than one operating mode is possible). These represent the strengths of the methodology.

Now for the limitations. One of them can be found by taking a closer look at the application factor representing environmental stresses. The value of the factor to be used for a specific application is tabulated by generalized class of service—"Ground, Benign;" "Ground, Fixed;" "Ground, Mobile;" "Space, Flight;" "Manpack;" "Naval, Sheltered;" and so on for a total of 20 categories. For a common type of transistor, the application factor ranges from a value of 1 for a "Ground, Benign" application to a value for 41 for a "Missile, Launch" application to 690 for a "Cannon, Launch" application. Obviously, the design engineer is required to employ subjective judgement in selecting the factor value that most closely represents his application. It must be acknowledged, however that each successive revision of the Handbook has provided finer gradations between factors so that less and less is being left to subjective judgement.

The other limitation stems from differences in the relative attention given in MIL-HDBK-217 to different component types. Solid-state devices and basic electronic components like resistors and capacitors receive extensive coverage, but from there on the analyst's road becomes progressively rockier. Electromechanical devices get short shrift, and non-electronic devices almost no attention at all. This is not surprising treatment in a reference document addressed to electronic equipment. Unfortunately, there is no comparable reference document for non-electronic equipment. Unless the component manufacturer provides reliability data on his product, you cannot find this information anywhere. Even when the manufacturer does provide data, it is frequently incomplete or expressed in nonstandard units, and almost always suspect.

The reason non-electronic equipment reliability is documented so poorly is that, with very few exceptions (notably, bearings), it is not feasible to standardize non-electronic components on an industry-wide basis. However, the designer does have other recourse. For many types of components, reliability standardization is not necessary, either because failure of the component would not affect the ability of the system to perform its mission (e.g., an elapsed operating time indicator on almost any type of equipment) or because the component is generally overdesigned for the stresses it will be subjected to (e.g., threaded fasteners, as selected in accordance with the manufacturer's instructions). For most components designed in-house, particularly mechanical components, a safety factor can be incorporated in the design such that the failure rate becomes insignificant and can be disregarded. However, for purchased non-electronic components whose failure would be critical with regard to reliability, and particularly for in-house designs for which a built-in safety factor is not feasible, the only source of quantitative reliability data is in-house testing.

In-house component testing conducted by most end-product manufacturers is generally limited to whatever is considered adequate for usage of that component in the product, and may consist of testing the product with the new components installed, rather than testing the components individually. Certainly, the results of this ad hoc testing are of limited value to the product designer planning to use the same components in an entirely different kind of application. Fortunately for the non-military designer, failure rates accurate to three decimal places are not usually necessary. As shown in Section 4.4, less precise data will frequently suffice.

1.9 USING RELIABILITY ANALYSIS IN DESIGN

Reliability analysis requires a significant investment of time and effort by the designer, but the payoffs are twofold: first, the analysis generates quantitative data that can be used to represent the reliability factor in cost/effectiveness studies directed toward optimizing the design; second, the reliability analysis generates information needed as an input to maintenance analyses (see Section 5.2).

Reliability data can be applied to design optimization in several ways. During the initial design stage—the stage in which the design concept is evolved—several alternate approaches to accomplishing the mission are formulated for evaluation. A preliminary analysis may disclose substantial differences in the reliability values estimated for the various candidate designs, thereby enabling the designer to weed out relatively unreliable alternatives. The analysis effort, in this instance, may consist of a comparison of parts counts for the alternatives, on the presumption that the reliability characteristics of the parts to be used, and the stresses to which these parts will be subjected, are essentially the same for all of the alternatives. Consequently, the only significant difference between the alternate designs is the number of parts required for each. However, if the stress levels or parts differ significantly from one alternative to another, it will be necessary to assign weighting factors accordingly. This can be done subjectively by the designer after he has performed a few failure modes analyses and matched the results against subsequently acquired field service data.

When the basic design has been selected, an initial failure modes analysis (FMA; see Chapter 4) should be prepared, in order to obtain an baseline reliability estimate for the product. If the designer already has a target reliability value in mind (and he should), the first FMA may indicate that the target has been exceed, or has not even been approached, or falls somewhere between these two extremes. If the target value has been exceeded, a safety margin is available to cover possible declines from the baseline reliability prediction as the original design concept evolves towards the detailed production version. If the prediction falls short of the target, the

the FMA will provide a list of failure sources that should be considered as candidates for reliability improvement.

The most obvious candidates for improvement are those components with relatively high failure rates. The nature of the improvement to be made depends on the kind of component involved, but component reliability improvements tend to fall into three general categories. The first category requires upgrading of the quality of the component; the second requires attenuation of the stresses causing the component to fail, and the third requires reinforcement of the component by redundancy. Advantages and disadvantages are associated with each of these methods.

Upgrading the quality of a high-failure-rate component is the simplest approach from the designer's standpoint, since impact on the design documentation may be limited to revising callouts on a parts list. However, a component of higher quality will almost certainly cost substantially more, and may entail longer procurement lead time (assuming that a higher quality component can be found).

Attenuation of failure-inducing stresses requires that the guilty stresses be identified and the design modified accordingly. If one of the culprits is an environmental stress, such as vibration or low ambient temperature, its effect on the reliability of vulnerable components can be reduced by incorporating the appropriate protective devices (e.g., vibration isolators, insulation, or thermostatically controlled heaters), as appropriate.

If another culprit is an operational stress, such as the power carried by an electrical resistor or the spring tension in a mechanical link, it is usually possible to reduce the stress level directly by modifying the circuit design. But, in those cases where the vulnerable component can be procured only as one of a family of components available in standard "sizes" (i.e., rated capacities), it may be preferable to replace the vulnerable component with the family member bearing the next higher rating. For example, substituting a two-watt resistor of the same resistance value for a one-watt resistor that dissipates one watt during normal operation will significantly reduce the component's failure rate by reducing its thermal stress at 100% rated load to the stress at only 50%. Logically enough, this technique is called "derating" the component. An alternative solution, in this case, would be to reduce the thermal stress on the resistor by installing a heat sink or other cooling provision.

Reinforcing high-failure-rate components with redundant components is a method employed only when the attendent penalties of additional weight, volume, and power consumption are acceptable. In most applications, redundancy also adds to the initial acquistion cost (purchase price) of the product, and the cost of maintaining it throughout its service life.

Incorporating a redundant component in a particular circuit of a product may necessitate changing the specification for the component. For example, if a resistor in an electrical circuit tends to fail frequently, another resistor can be connected in parallel with it so that, if either resistor fails, the other will still complete the circuit. But, because the current flow is now divided between two resistors, the resistance value originally specified will have to be increased, if the designer wishes to avoid changing the total current flowing in the circuit. He must also be aware that, if one of the redundant pairs fails, the effective resistance within the circuit will change from the equivalent resistance of the parallel pair to that of the surviving resistor. Different circuit configurations call for different redundancy arrangements. The technique for designing redundancy into a circuit is discussed in Chapter 7.

For airborne equipment applications, where the physical penalties associated with redundancy are particularly undesirable, its employment is avoided whenever possible. Conversely, for the Federal Aviation Administration's Air Traffic Control and Navigation System, the need for extremely high reliability levels dictated by safety considerations, coupled with the relatively low penalties incurred by fixed-location ground-based stations, make redundancy the preferred design option. A typical radar transmitter/receiver station employs two completely separate circuits operating in parallel (designated as Channels A and B), except for a single antenna assembly fed by either channel. When one channel is on the air, the other channel is in the standby mode, or being checked out, or undergoing repair. If the operating channel should suddenly malfunction, the other channel will immediately be switched in to replace it. Consequently, 100% redundancy of the operating channel (except for the antenna) exists at all times.

From the designer's standpoint, the reliability story can be summarized as follows:

1. For a given design, the cost increases (usually exponentially) as the reliability is improved.
2. The designer's goal is a product that will achieve an in-service reliability level satisfying the customer while limiting the impact on manufacturing cost to an acceptable level.
3. The designer can control the product's reliability by selecting an appropriate combination of: design concept alternatives; detailed circuit configuration; part quality levels; and redundancy. But to make the selection optimal, he must

have some means of determining the reliability offered by any given combination of these factors, and the effect of varying any one of them.

4. The designer can utilize the procedures in Chapter 4 to compare the reliabilities predicted for the design alternatives, and to assess the effects of modifying circuits, changing quality levels, or introducing redundancies. How close he comes to the absolutely optimal combination depends on how much time he wants to spend on the evaluation, but he can get very good results with only moderate effort.

APPENDIX

Before 1960, you probably could not have found a single book on reliability—at least, I couldn't. But beginning with the decade of the 'sixties, reliability texts have rolled off the printing presses in a steady procession, differing somewhat in where they place emphasis but generally agreeing on fundamentals and virtually always exhibiting a strong mathematical orientation.

One of the first comprehensive publications on reliability was written by Igor Bazovsky (1), and it became a standard reference for reliability analysts for many years. His book is one of the few that mentions the pioneering contributions of Lusser and Pieruschka. I had the privilege of working with Igor briefly in 1962, and recall with pleasure an informal lecture he once gave on mission definition.

Another early publication, that of Lloyd and Lipow (2) has been updated to suit the needs of contemporary reliability analysts. Anyone comparing a formal text like that of Lloyd and Lipow with the book you are now holding will discover that there is a much greater store of mathematical tools available to the dedicated analyst than what you find here. For example, what we have called the fundamental reliability equation is based on just one of several mathematical models of failure distribution, this one derived from an expression designated formally as the *exponential probability density function*, and is stated like this:

$$f(t) = \lambda e^{-\lambda t}$$

For certain applications, particularly those involving mechanical components, the so-called two-parameter *Weibull distribution* is considered a more suitable model—it looks like this:

$$f(t) = d\lambda t^{\alpha-1} e^{-\lambda t \alpha}$$

The Weibull distribution appears far more complex than the exponential distribution but, the latter is actually on a special case of the former. If you set α in the Weibull expression equal to unity, the expression is simplified to become the exponential probability density distribution, which is accurate enough for our purposes.

One of the most useful references for reliability analysts dealing with electronic equipment is Military Handbook MIL-HDBK-217 (3), which came in for considerable discussion in the main text of Chapter 1. The current revision of the Handbook provides basic failure data and application factors based on a vast body of laboratory test results and field experience. However, some of the supplementary discussion on probability—based prediction methodology that had appeared in earlier issues has been transferred to a separate publication, Military Standard MIL-STD-756 (4), where it now appears in expanded form. (Military documentation of this kind is not published for the public but is readily available to organizations working under contract to an agency of the Department of Defense.) If you need a rigorous discussion of the probabilistic foundation of reliability, try Siddall's new book (5), which is directed toward risk assessment in design. His Chapter 9 treats the prediction of failure mode probabilities, Chapter 11 addresses reliability theory, and Chapter 12 the application of reliability concepts to structural members.

Dummer and Winton (6) provide a general introduction to reliability, but their presentation may be found somewhat outmoded by now. Another early publication is that of Pieruschka (7) himself. More recent are books by Greene and Bourne (8) and Siewiorek and Swarz (9). The mathematics of reliability is discussed by Amstadter (10). If you need more information on testing to improve reliability, you might want to look at Gilmore and Schwartz (11). Carter (12) concentrates on the reliability of mechanical systems.

REFERENCES

1. I. Bazovsky, *Reliability Theory and Practice*, Prentice-Hall, Inc., Englewood Cliffs, N.J., 1972.
2. D. Lloyd and M. Lipow, *Reliability: Management, Methods, and Mathematics*, published by the authors, Redondo Beach, Calif., 2nd Ed., 1977.
3. Military Standardization Handbook, MIL-HDBK-217, *Reliability Prediction of Electronic Equipment*, U.S. Department of Defense, Rivision D, 1982.

4. Military Standard, MIL-STD-756, *Reliability Modeling and Prediction*, U.S. Dept. of Defense, Revision B, 1981.
5. J. Siddall, *Probabilistic Engineering Design: Principles and Applications*, Marcel Dekker, Inc., New York, 1983.
6. G. W. A. Dummer and R. C. Winton, *An Elementary Guide to Reliability*, Pergamon Press, New York, 1968.
7. E. Pieruschka, *Principles of Reliability*, Prentice-Hall, Englewood Cliffs, N.J., 1963.
8. A. E. Greene and A. J. Bourne, *Reliability Technology*, Wiley-Interscience, New York, 1972.
9. D. P. Siewiorek and R. S. Swarz, *The Theory and Practice of Reliable System Design*, Digital Press, Bedford, Mass., 1982.
10. B. L. Amstadter, *Reliability Mathematics: Fundamentals, Practices, Procedures*, McGraw-Hill, New York, 1971.
11. H. L. Gilmore and H. C. Schwartz, *Integrated Product Testing and Evaluation: A Systems Approach to Improve Reliability and Quality*, Wiley-Interscience, 1969.
12. A. D. S. Carter, *Mechanical Reliability*, Wiley, New York, 1972.

2 Maintainability

2.1 INTUITIVE MAINTAINABILITY CRITERIA

The roots of the maintainability discipline, like those of reliability, can be traced back to the beginnings of the Industrial Revolution. Unlike the reliability discipline, however, awareness of the need for a maintainability discipline did not materialize until the eve of the Second World War—but then it zoomed to keep pace with the newly-emerged high-technology industries. Again, unlike reliability, maintainability never suffered a temporary loss of credibility because of exaggerated claims made for its capabilities.

Maintainability began as a set of qualitative guidelines formulated in response to the demands of mechanics who maintained the designers' products after they were placed in service. Mathematical underpinnings were subsequently furnished by theoreticians seeking to apply reliability prediction techniques to this allied field.

The following may be regarded as a common-sense definition:

> *Maintainability* is that element of product design concerned with assuring that the ability of the product to perform satisfactorily can be sustained throughout its intended useful life span with minimum expenditure of money and effort.

Even in the early days of the Industrial Revolution, every good product designer had intuitively developed a set of design guidelines intended to enhance the maintainability of the product. As mass-production techniques became commonplace, designers tended more and more to stress producibility over maintainability, but it appears that some of the design practices that facilitate manufacturing also favor maintainability. Most of the fundamental maintainability principles described in the sections below obviously originated as efforts to accelerate production lines.

2.1.1 Standardization

Standardization enforces the conformance of commonly used parts and assemblies (e.g., fasteners, bearings, plumbing system components, auto wheels, electrical connectors) to generally accepted design standards for configuration, dimensional tolerances, performance ratings, and other fundamental design attributes. The objectives of standardization are to:

1. Assure compatibility between mating parts, when replacing a faulty unit, and between the product and the common tools, test equipment, and facilities used for its maintenance;
2. minimize the number of different spare parts that must be stocked for maintenance support.

2.1.2 Modularization

Modularization enforces conformance of assembly configurations to dimensional standards based on modular "building-block" units of standardized size, shape, and interface locations (i.e., locations for mating attachment or mounting points and input/output line connectors), in order to simplify maintenance tasks by enabling the use of standardized assembly/disassembly procedures.

2.1.3 Functional Packaging

Functional packaging locates all components of a circuit performing a given function in the same package. (A package, in this context, is a unit that is readily removable and replaceable as an entity.) The objective of functional packaging is to expedite repair of a faulty system. If a given circuit is entirely packaged in a single replaceable unit, the symptoms manifested by the failure of that circuit identify the faulty unit. Replacement of that single faulty unit completely corrects the failure.

2.1.4 Interchangeability

Interchangeability controls dimensional and functional tolerances of manufactured parts and assemblies to assure that a faulty item can be replaced in the field with no physical rework required for achieving a physical fit, and with a minimum of adjustments needed for achieving proper functioning.

2.1.5 Accessibility

Accessibility controls the spatial arrangement of parts and assemblies within a piece of equipment so that each of these items is readily accessible for replacement or repair in-place. Accessibility can be facilitated by provision of carefully located hinged doors, removable panels, or sliding shelves or brackets, as applicable. For the purposes of maintainability analysis, accessibility is also considered to encompass provisions for facilitating removal and replacement of faulty items, such as quick-disconnect terminations on electrical, hydraulic, and other lines connected to items that may require replacement. Evaluation of the relative accessibility of each component of a given design must take into consideration the physical limitations of the maintenance worker (human factors), and whether other items must first be removed in order to gain access to a specific item.

2.1.6 Malfunction Annuciation

Malfunction annuciation provides for indicating to the operator that the equipment is malfunctioning, in those cases where a malfunction is not readily evident. Warning indicator lights on the dashboard of your automobile are intended to alert you to malfunctions in your engine cooling system, engine oil system, or electrical system so that you can stop before losing power or damaging the engine.

2.1.7 Fault Isolation

Fault Isolation assures that an equipment malfunction can be traced to the faulty part of assembly requiring replacement or in-place repair, even if supplementary hardware must be provided solely for that purpose. Fault-isolation provisions can be of various types.

Perhaps the simplest are readily accessible connectors or probe receptacles ("test points") for plugging in external test equipment that can determine whether certain components are operating properly or should be replaced. At the other end of the spectrum are subsystems integrated into the product's primary system that can run complete checkout programs and indicate which, if any, components are faulty. Integrated malfunction detection and fault isolation subsystems are commonly known as *BITE (built-in test equipment)*.

2.1.8 Identification

Identification utilizes engraving, marking, or labelling for quick location of parts or assemblies upon which preventive or corrective maintenance may have to be performed; for interconnecting lines, color coding is also used. Standard military practice calls for the following markings, in addition to part numbers on piece-parts and assemblies, in order to enhance maintainability:

1. *Serial numbers* uniquely identifying each production unit of a critical piece-part, assembly, or equipment item bearing a given part number, to facilitate the recording of that item's operating and repair history throughout its service life.
2. *Reference designations*, which are complementary designations for replaceable components and their corresponding installation locations. Reference designations are desirable for plug-in components such as printed-circuit-board assemblies, particularly when the same part-numbered component is used in more than one place in the product. (Another desireable maintainability feature for plug-in components is the incorporation of some form of mechanical interference provision for preventing inadvertent insertion of the component in the wrong receptacle or in the wrong position in the correct receptacle. Such provisions are sometimes called "*indexing*.")
3. Conductor identification numbers marked on lengthy wires and cables at regular intervals, and complementary designations for mating connector halves (e.g., J1 on the female connector half of Conductor Number 1, and P1 on the male half).

Our checklist is a compilation of qualitative maintainability principles that translate directly into good design practice. However, it is also possible to express and predict the relative maintainability of a given design in quantitative terms, as we shall discover next.

2.2 EMERGENCE OF THE MAINTAINABILITY DISCIPLINE

Maintainability, as an acknowledged engineering discipline, came into existence because the military services had come to recognize that, at any point in time, a large proportion of their equipment was unavailable for use because it was either undergoing or awaiting repair. A dramatic example was afforded by the U.S. Army's Eighth Air Force based in England during World War II—in one instance it was found that only 30% of the heavy bombers stationed at one field were in mission-ready condition when needed, and the situation at other fields was not much better. This intolerable unavailability of equipment was traced to two general causes:

1. failure to design the equipment and its supporting hardware as an integrated system from the maintenance standpoint. Designers had simply not taken into account the conditions under which the user had to perform maintenance in the field.
2. ineffective logistics planning and planning implementation for maintenance resources. Tools, test equipment, special facilities, spare parts, properly trained technicians, and technical manuals were not readily available where and when needed.

The logistics problem gave rise to a whole separate discipline which the military has designated as integrated logistic support *(ILS)*—we'll touch on logistics when we discuss maintenance planning in Chapter 6. As for the design problem, its solution required finding some quantitative measure of effectiveness for evaluating and comparing alternative proposed maintenance improvements. The analytic work being done in the reliability field suggested that the maintainability measurement unit be expressed in terms of time, and be analagous to the MTBF unit for expressing the inherent reliability of a given design. And so, the military came to adopt *mean time to repair (MTTR)* as the unit for expressing the maintainability inherent in a given design. The total calendar life of a product can be divided into periods of operation (or standby mode available for operation), separated by periods when the product is inoperative because it is awaiting or undergoing repair.

2.3 PREDICTING AND MEASURING MEAN-TIME-TO-REPAIR (MTTR)

The calculations to be made in predicting the MTTR of a product are not as simple as those required for predicting its MTBF. In order to calculate MTTR, the time required for performing each anticipated repair task must be weighted (multiplied) by the relative frequency with which the need for that task is expected to arise during the service life of the product. Since MTTR represents the *mean* time to restore a malfunctioning system to proper operation, the undesirable effect of a repair procedure that takes a comparatively long time to accomplish but does not have to be performed very often can be offset by another procedure that has to be done more frequently but can be completed quickly.

In determining MTTR, the relationship between repair task times and the relative frequencies with which they must be performed can be illustrated by using a set of hypothetical repair frequency/ time measurements compiled for Repair Tasks A through Z during a hypothetical product's useful life. Table 2-1 lists the time to perform each task, the number of times that the task was performed during the service period (frequency of occurrence), the aggregate (cumulative) time expended during the service period in performing that task (individual task performance time multiplied by frequency of occurrence), the relative frequency of occurrence of that task, and the percentile (cumulative percentage) represented by that task.

The simplest way to graph a set of statistical data for analysis is in the form of a histogram. In the histogram shown in Figure 2-1, the horizontal axis represents corrective repair task time, and is divided into segments of six minutes each. Each number shown along the horizontal axis denotes the midpoint of the corresponding segment; any given segment spans the horizontal distance from a point three minutes less than the midpoint value to a point three minutes greater than the midpoint value. The midpoint approximates the mean of the task times falling within the segment; the briefer the segment the more accurate we can expect the approximation to be.

Every task time listed in Table 2-1 that falls within the boundaries of a given segment is represented on the histogram by one vertical unit, thereby forming a column whose height indicates the frequency with which tasks of that duration were performed during a selected service period. Thus, the first column on the left indicates that 10 of the recorded task occurrences range from three to nine minutes duration (i.e., six minutes, plus or minus three), and may be assumed to average approximately six minutes per task; the greater the number of tasks in the column, the more accurate the approximation. (Task times ranging from zero to three minutes are presumed to fall into the vertical column whose midepoint is zero minutes; since the task time data in Table 2-1 is expressed in

Table 2.1 Repair Task Time Data (Hypothetical)

Task	Time to perform (hours)	Frequency of occurrence	Aggregate task time (hours)	Relative frequency of occurrence (percent)	Percentile
A	0.1	2	0.2	1.7	1.7
B	0.1	3	0.3	2.5	4.2
C	0.1	3	0.3	2.5	6.7
D	0.1	2	0.2	1.7	8.3
E	0.2	2	0.4	1.7	10.0
F	0.2	2	0.4	1.7	11.7
G	0.2	4	0.8	3.3	15.0
H	0.2	5	1.0	4.2	19.2
I	0.2	5	1.0	4.2	23.3
J	0.3	4	1.2	3.3	26.7
K	0.3	4	1.2	3.3	30.0
L	0.3	5	1.5	4.2	34.2
M	0.3	5	1.5	4.2	38.3
N	0.3	5	1.5	4.2	42.5
O	0.4	4	1.6	3.3	45.8
P	0.4	4	1.6	3.3	49.2
Q	0.4	5	2.0	4.2	53.3
R	0.5	4	2.0	3.3	56.7
S	0.5	5	2.5	4.2	60.8
T	0.5	5	2.5	4.2	65.0
U	0.5	6	3.0	5.0	70.0
V	0.5	8	4.0	6.7	76.7
W	0.6	7	4.3	5.8	82.5
X	0.8	10	8.0	8.3	90.8
Y	1.5	9	13.5	7.5	98.3
Z	1.6	2	3.2	1.7	100
Totals		120	59.6	100.0	

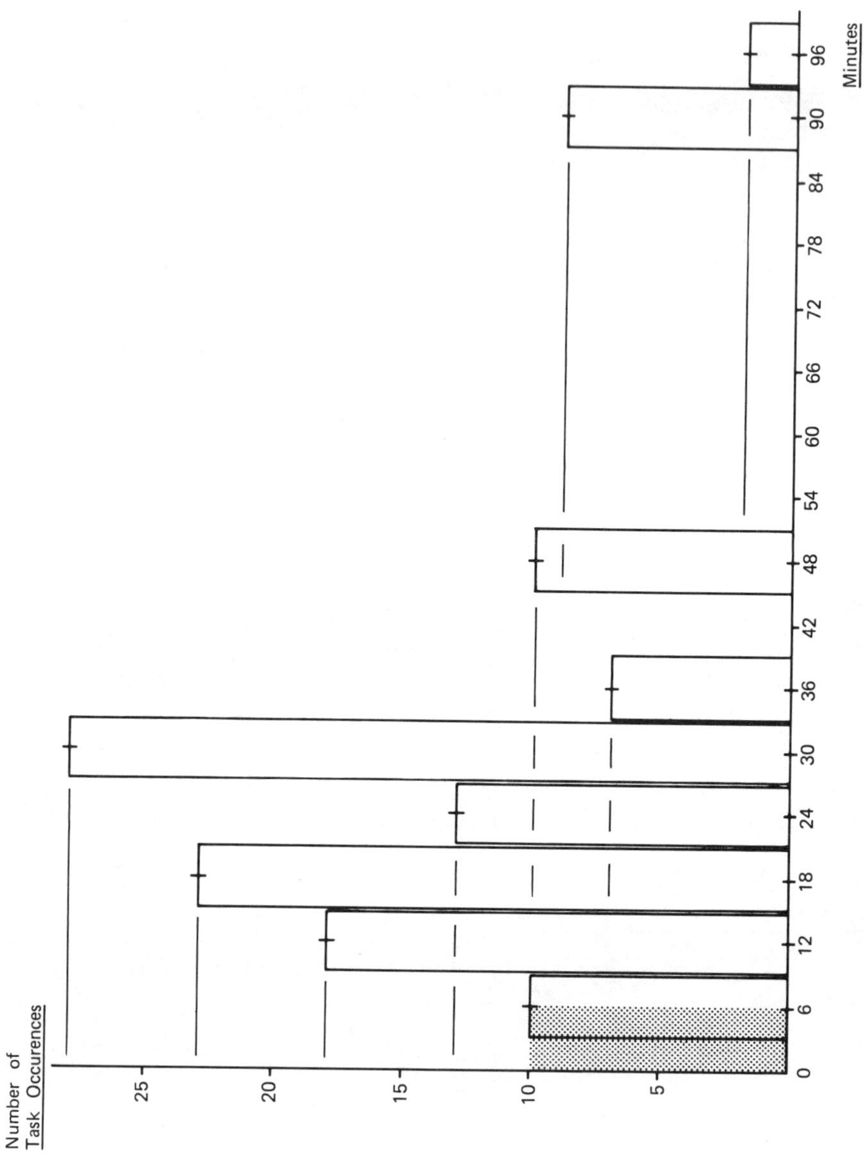

Figure 2.1 Distribution of corrective maintenance task times.

decimal values to the nearest tenth of an hour, and none of the task times is zero, we have no zero time column on our histogram.)

In Figure 2.1, the shaded area between the vertical axis and the vertical line passing through the midpoint of the six-minute task time column represents the aggregate time expended in performing the ten tasks that fall into that column. Ten performances of tasks averaging six minutes each account for an area representing an aggregate of 60 minutes. Similarly, the area between the vertical axis and the vertical midpoint line of each of the other columns represents aggregate time consumed by the tasks that fall into that column. Obviously, the aggregate task time area represented by each column is determined by: (a) the distance of the midpoint of that colunn to the right of the vertical axis; and (b) the height of the column.

The total repair time expended in maintaining the product throughout the designated service period is obtained by summing the aggregate areas of all the columns. The MTTR for the product is found by dividing the total number of tasks performed during the service period into this sum. The MTTR can be regarded as the mean of the total time expended in performing all of the tasks, and, therefore, the mean time required for restoring a malfunctioning product to proper operation.

The relationship between the MTBF (reliability) and MTTR (maintainability) units can be seen by considering that the service life of a product consists of operating and standby periods (up-time) separated by periods when the product is awaiting or undergoing repair (down-time). Comparison of a product's MTBF and MTTR indicates the relative magnitudes of the up-time and down-time that can be expected of it. However, MTBF measures only that portion of the up-time during which the product is actually operating. Similarly, MTTR measures only that portion of down-time expended in actively performing repairs; the time lost while awaiting the arrival of spare parts or tools or technicians or completion of preliminary paperwork is not influenced by the product's design, and must therefore be charged to "administrative waiting time," rather than to MTTR.

Just as MTBF predictions provide a quantitative means for comparing the reliabilities of alternative designs, so MTTR predictions serve as a comparison of the relative maintainabilities of alternative designs, so MTTR predictions serve as a comparison of the relative maintainabilities of alternative designs. Just as MTBF predictions can be used in managing the reliability characteristics of a product design during development, so MTTR predictions can be used in managing its maintainability characteristics. However, the military services have found that specifying MTTR alone will not prevent the occasional occurrence of extremely lengthy down-times; i.e., long periods when the product is out of service undergoing repair. This occurs when excessively time-consuming tasks are offset by more frequent short-duration tasks, thereby permitting the MTTR specification to be satisfied even though the longest out-of-service periods exceed tolerable limits.

This point can be illustrated using the data listed in Table 2-1. Supposing that a maximum allowable MTTR of 0.5 hours (30 minutes) had been specified by the customer for the product analyzed in Table 2-1. Task Z, which was performed twice during the service period, required 1.6 hours (96 minutes) each time. But the extermely lengthy Task Z can be offset by Tasks A through C, none of which takes more than 0.3 hours (18 minutes); if repair of the product consisted only of these four types of tasks, the MTTR of the product would be only 0.4 hours (24 minutes), even though two of the tasks each took four times as long as the MTTR. For the aggregate of the tasks from A through Z, the product's MTTR is 0.497 hours, slightly less than 30 minutes. If it were necessary to limit down-times to no more than 60 minutes, the design in our hypothetical example would not be acceptable, since two down-times of 96 minutes each are seen to occur, even though the 30-minute MTTR requirement has been satisfied.

2.4 PREDICTING AND MEASURING MAXIMUM REPAIR TIME (M/max/ct)

To close the loophole in controlling the maintainability of the design, a second quantitative maintainability unit was devised, designated as M/max/ct. To explain this unit, let us plot the hypothetical repair times as a cumulative distribution curve (Figure 2-2), which is developed in the following manner: First, the compiled task data is arranged in order of increasing task completion times, and grouped into time span increments of six minutes (0.1 hour) duration each, just as we did for the histogram of Figure 2-1. Second, a series of horizontal line segments is plotted from left to right, beginning at the vertical axis. The first of these segments represents the task grouping with the lowest completion time increment. Its horizontal length is 10 task occurrences (from the zero percentile to the 8.33 percentile, measured along the horizontal axis). The height of this segment, measured against the vertical

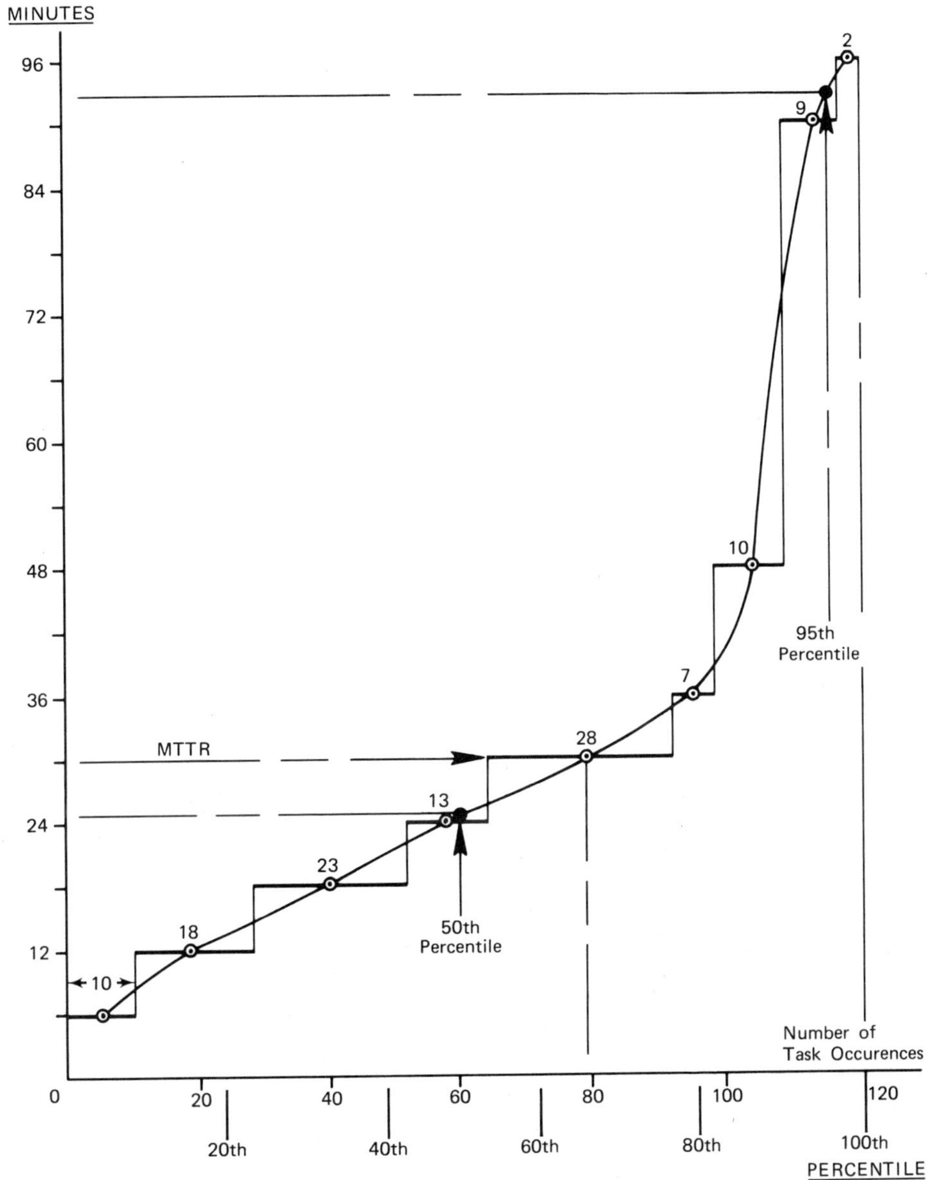

Figure 2.2 Cumulative distribution of corrective maintenance task times.

axis (performance time), is six minutes. Each successive segment represents the task grouping with the next higher completion time increment.

By connecting the midpoints of each of the segments, and smoothing the resulting line to eliminate angular transition sections, a continuous curve is derived that simulates the plot of a very large population of measured task times. Every point on this curve represents the maximum task completion time for that portion of the task population to the left of the point on the curve. The horizontal coordinate of any point on the curve designates the corresponding percentile of the population.

Note that the points along the cumulative distribution curve labeled "MTTR" and "50th percentile" do not coincide. The 50th percentile point indicates that 50 percent of the task population (60 task performances, for our example) is to the left of this point; the task completion time for the 50th percentile is slightly greater than 24 minutes. The MTTR, which is equivalent to the mean task completion time, is 29.8 minutes. The reason for the difference between the two can be seen from the histogram in Figure 2-1—most of the task occurrences have very short completion times but a few have very long completion times. If the occurrences had been grouped symetrically about the 50th percentile, as in the normal ("bell-shaped") distribution familiar to students of statistics, the MTTR point would have coincided with the 50th percentile point. But the distribution in our example is skewed to the left, displacing the MTTR from the 50th percentile.

But what has all this to do with M/max/ct? Simply that *M/max/ct* is defined as the maximum time allowed for completing every task below a stated percentile, which is usually taken to be the 90th or 95th percentile. (In this book, we will always use the 95th percentile.) The presumption here is that a specifically limited margin of extra-long out-of-service periods is tolerable.

On the histogram (Figure 2-1), the 95th percentile task can readily be traced to the 90-minute column, but might actually fall anywhere from 87 to 93 minutes within that column. On the cumulative distribution curve (Figure 2-2), the 95th percentile task appears to be at about the 93-minute point. Obviously the latter is only a very rough approximation because of the method of constructing the cumulative distribution curve to represent a large population. This is a perfectly respectable statistical technique but, in this case, constitutes a distortion of the data (Table 2-1), which had been formulated to simulate field practice, where there is a tendency to lump time measurements into modular time spans for convenience. (During field operations, it's very difficult to judge precisely the moment when a repair task begins or ends.)

The M/max/ct for a given product can be approximated by either of the methods just described for plotting the histogram and the cumulative distribution curve of repair task times or, when a precise value is required, can be calculated from the following rather messy equations:

$$\text{M/max/ct} = \text{antilog}_e\ (A + B)$$

$$A = \frac{\sum_{i=1}^{N_c} \ln X_{ci}}{N_c}$$

$$B = \psi \sqrt{\frac{\sum_{i=1}^{N_c} (\ln X_{ci})^2 - \left[\left(\sum_{i=1}^{N_c} \ln X_{ci}\right)^2 / N_c\right]}{N_c - 1}}$$

where

N_c is the total number of repiar tasks performed, X_{ci} is the time expended in performing the 'i'th task, and ψ is a coefficient whose value depends on the percentile chosen as the maximum repair time boundary (for 90th percentile, $\psi = 1.282$; for 95th percentile, $\psi = 1.645$).

2.5 A COST-EFFECTIVENESS MEASURE OF MAINTAINABILITY

For certain types of applications, additional quantitative maintainability measurements have been devised, such as maintenance labor-hours per operating hour. These units enable a more detailed management of labor costs, which constitute the greatest single maintenance expense factor for most products. Although MTTR and M/max/ct address time expended in performing maintenance (clock hours), they do not consider the number of persons performing a given reapir task concurrently; nor do they consider labor expended in performing maintenance tasks that do not involve repairs; i.e., preventive maintenance, as described in the next section.

2.6 MAINTAINABILITY VERSUS MAINTENANCE

The achievement of optimal maintainability requires that the design demonstrate maximum compatibility with the user's maintenance capabilities and procedures. Those procedures have been developed by the user over a considerable period of time, and are not readily changed. Furthermore, they are probably patterned after common practice, which is generalized to serve a variety of equipment types. The designer must, therefore, tailor the product to accommodate the user's maintenance resources and practices.

Maintenance consists of two general types of tasks:

1. *Preventive maintenance*, which comprises and is limited to those tasks performed in order to defer or prevent an anticipated failure occurrence. Preventive maintenance thus serves to extend actual product reliability beyond the MTBF value that would be predicted on the premise that every component is allowed to operate until it fails.
2. *Corrective maintenance*, which comprises and is limited to those tasks performed in order to restore the product to acceptable operating condition after a failure has occurred.

Preventive maintenance can be justified on a cost-effectiveness basis only when it eliminates more costly corrective maintenance. One cost advantage preventive maintenance enjoys over corrective maintenance is that preventive maintenance can usually be scheduled for times when the product will not be in operation; the time expended in performing corrective maintenance must be treated as a loss of capabilities already paid for whenever it renders the product unavailable for intended service. An additional justification, but one that cannot easily be quantified, is the avoidance of inconveniences or hazards resulting from certain types of failures. An auto engine that fails during a trip is an example of the former, faulty brakes exemplify the latter.

2.6.1 Preventive Maintenance

Preventive maintenance tasks are of five kinds:

Servicing

Tasks that must be performed periodically to assure that an equipment item will remain in proper operating condition. Servicing includes (but is not necessarily limited to):

1. Replenishment of consumables such as fuel, and of depletables such as lubricants, lead-acid battery electrolyte, coolant fluids, and air (in pneumatic tires)
2. Minor adjustments and alignments
3. Cleaning of permanent filters and other critical surfaces subject to contamination
4. Replacement of expendable filters and other short-lived items

A servicing task consists of an inspection to determine the extent of work required, followed by the performance of that work.

Condition-Monitoring

Surveillance of the operation of the product (usually by the operator) to detect abnormalities indicative of an impending failure or of a failure that has already occurred.

Condition Assessment and On-Condition Remedial Action

Condition assessment is evaluation of the physical condition of product components known to be subject to wear or other forms of deterioration; condition assessment is performed during scheduled inspections of the product. On the basis of this evaluation, it is decided whether the item can safely be permitted to remain in use as-is until the next scheduled inspection or should be subjected to remedial action. The remedial action can consist of replacing a suspect component, or of adjusting, aligning, calibrating, refurbishing, renovating, or repairing the equipment in-place. (These terms are defined a little further on.)

Typically, condition assessment is performed on: surfaces that wear out or corrode (e.g., bearings, gear teeth, tire tread, DC motor commutators and brushes); mechanisms subject to misalignment or excessive play (e.g., mechanical operating controls); items that accumulate contaminants interfering with operation (e.g., radiator cores through which fluid coolant circulates). For electronic equipment, however, the bulk of condition-assessment tasks may consists of measurements of performance parameter drift, to assess the need for adjustment, alignment, or calibration.

If component replacement is found to be necessary, it is designated as *on-condition* remedial action, as distinguished from scheduled replacement, as defined below. If in-place repair is found necessary, it too is designated as on-condition remedial action.

Methods used for condition assessment include visual examination (with or without an optical instrument); measurement of physical and/or functional attributes, using the appropriate test equipment;

and various special nondestructive test techniques. Condition assessment is performed only for attributes subject to degradation that ultimately leads to failure. (It should be noted that certain condition-monitoring methods that go beyond GO/NO-GO indication also constitute condition assessment, making scheduled inspection unnecessary for those conditions.)

Verification of Hidden Functions

Hidden functions are functions not exercised during normal operation of the product and, therefore, not subject to condition-monitoring. The circuit or subsystem performing the hidden function must be exercised periodically to verify that it is still capable of performing the function.

Typically, hidden functions are emergency or redundant functions. A circuit-breaker performs an emergency function; a backup power supply performs a redundant function that enhances the reliability of the product. Both are hidden functions that require special techniques and, where feasible built-in test provisions for verifying that the capability is available when needed. If at all possible, verification of the function should exercise the corresponding circuit or subsystem in every one of its operating modes. If this is not feasible, some sort of indirect test method is needed. Unfortunately, indirect methods do not always provide absolute verification. In the case of a fuse, the function cannot be exercised directly, and the only economically-feasible indirect methods are nondestructive examination and verification of electrical continuity.

Scheduled Replacement

Scheduled replacement applies to components that meet all of the following criteria:

1. Failure of the component would endanger personnel or equipment, or would reduce the operational availability of the product (see Chapter 3) below the minimum acceptable level.
2. The failure mode is not suitable for condition assessment, which is the preferred preventive maintenance method.
3. The useful life span of the component is substantially less than the intended operating life of the product, and can be determined relatively accurately (in other words, a limited-life component.)

The proper selection and scheduling of preventive maintenance tasks requires a methodology all its own. This need has been met by "Reliability Centered Maintenance," which is described in Chapter 4.

2.6.2 Corrective Maintenance

Corrective maintenance tasks are generally accomplished in four steps: (1) fault detection (detection of a system malfunction occurrence); (2) fault isolation (tracing of the malfunction cause to a lower-assembly-level failure source); (4) fault elimination; and (4) verification that the fault has been eliminated.

Fault elimination may be accomplished in any one of three ways, depending on the nature of the fault:

1. *Adjust/align/calibrate.* Restoring the product to acceptable operation by adjustment, alignment or calibration of a circuit or component controlling a critical performance parameter.

To *Adjust* is to vary the setting of a controller until the performance parameter that it controls is brought into an acceptable state or falls within an acceptable range; e.g., to adjust the carburetor of an automobile engine for minimum fuel consumption at cruising speed.

To *align* is to bring a subsystem or circuit to a designated position(s); e.g., to align the front suspension of an automobile such that both front wheels assume the proper attitude, with regard to the road, for all positions of the steering wheel.

To *calibrate* is to verify, for a variable controller, the actual performance value of the parameter controlled at a number of predesignated controller settings, correcting any discrepancies found; e.g., the shutter speeds of a camera at the various settings marked on the shutter speed control.

2. *Rework.* Physically altering a faulty component to restore it to its originally designed structural or functional condition, as shown on the pertinent engineering drawings, or to another condition deemed acceptable. Commonly used rework methods include welding, soldering, patching, splicing, bonding, and reinforcing. Rework is particularly applicable to the correction of failures originating in faulty interconnections utilizing media such as wiring, piping, hoses, mechanical linkages, or optical fibers.

3. *Remove-and-replace (R&R).* Restoring a malfunctioning product to proper operation by removing the faulty subordinate component (the failure source) and replacing it with a properly functioning spare component, with no rework involved.

The general term, maintenance, as used in this book excludes two kinds of tasks that are intended to prolong the effective operating life of an equipment item. The excluded tasks are:

1. *Refurbishment.* Cleaning, painting, and related tasks performed for the purpose of restoring surfaces degraded by abrasive

or corrosive agents in the atmosphere, or by scratches, nicks, or gouges received during in-service handling. Refurbishment can be thought of as addressing "cosmetic blemishes" rather than degradation of functional capabilities or useful life.

2. *Renovation.* Restoration of some or all of an item's operating life span after it had been expended by prior usage. Renovation generally entails restoration of worn or abraded elements of the item's basic structure, as well as replacement of worn, damaged, or degraded removable parts. Renovation encompasses modifications of the original design, such as enlarging a worn journal bore and inserting an oversized bushing, or machining the commutator of an electrical motor to restore smooth electrical contact surfaces. The objective of renovation is to reduce life-cycle maintenance support costs by salvaging components whose aggregate restoration costs are significantly less than current replacement costs. Rework, on the other hand, is intended solely to correct a fault, rather than normal wear or other deterioration. (The term, renovation, is used in place of the more familiar term, overhaul, in order to avoid conflicting interpretations of the latter in common usage.)

2.7 DESIGNING FOR MAINTAINABILITY

This chapter opened with the idea that maintainability had begun as a common-sense approach taken by designers at the behest of the end-user; their objective was to assure that the product could be maintained throughout its useful life at reasonable expense and without any particular difficulty. The approach consisted of evaluating the design of the product against a checklist of qualitative maintainability principles. More recently a concept has been adapted from the reliability field to provide ways of expressing quantitatively the maintainability inherent in the design. Chapter 8 describes methods for calculating circuit and system maintainability from component data compiled utilizing the procedure set forth in Chapter 4. But the question must be asked—do these quantitative expressions provide a capability to the designer beyond that afforded by qualitative principles? The answer is yes, but only if the numbers are utilized intelligently. It must also be stressed that quantitative maintainability measurement methods supplement the qualitative principles but do not supersede them.

In evaluating maintainability analysis results, it should never be forgotten that MTTR and M/max/ct predictions are only estimates based on prior experience and the analyst's judgment—they are not the last word. Furthermore, these predictions represent the capabilities of the product only when maintained under ideal support conditions, conditions rarely even approached in the real world—a design with a demonstrated "hands on" MTTR of less than 30 minutes may be very impressive—but what if it takes two weeks to get the necessary spare parts to the user after the equipment broke down?

However, even after accounting for these practical disclaimers, a solid foundation of real worth remains. The value of quantitative maintainability analysis lies in how it can be used in product design and maintenance planning; the latter is covered in Chapter 6. Design modifications proposed on the basis of analysis results for improving maintainability can be subjected to cost-effectiveness evaluation, comparing the one-time cost of making the design change with the end-user's ongoing savings in maintenance expenditures over the lifetime of the product.

Because the designer must invariably make compromises to achieve the optimum balance between various design factors—performance, weight, volume, energy consumption, reliability, maintainability, and so on—trade-off studies are necessary. These studies require that the design parameters be stated in quantitative terms that can be expressed in a common unit of measurement, such as equipment operating and non-operating hours, and then converted into dollar equivalents. MTTR and M/max/ct constitute equipment non-operating hour units developed for this purpose. Maintenance labor hours per operating hour is another measure of maintainability convertible to dollars. For trade-off analysis purposes, the complete life-cycle support cost is the sum of the dollars spent on maintenance plus the dollars lost whenever the product cannot be used because it is undergoing repairs. Chapter 6 identifies the resources that must be provided (and costed) for maintenance; Chapter 3 deals with measuring lost opportunity for use.

However, life-cycle support cost may not be the only concern of the designer when evaluating the maintainability of the product. For many applications, lengthy repair sessions can have undesirable consequences ranging from inconvenience to deadly hazard. The repair task times compiled as part of the procedures described in Chapter 4 for preventive maintenance will enable the designer to identify the tasks that must be shortened by making the appropriate design modification.

APPENDIX

Books dedicated specifically to maintainability are not as numerous as those on reliability, but one comprehensive reference is that of Blanchard and Lowery (1). Another source is Cunningham and Cox (2). Maintainability is also discussed in books on logistic planning (see appendix to Chapter 6).

The equations for M/max/ct in the main text of this chapter may be found in Section 2.6 (Pages 3-12 & 3-13) of Military Handbook MIL-HDBK-472 (3) and in Test Method 9 of Military Standard MIL-STD-471 (4).

Military Handbook MIL-HDBK-472 describes four different maintainability prediction procedures; the approach taken in this chapter for calculating MTTR parallels that presented in the Handbook as Procedure II, which is applicable to all kinds of systems and equipment.

REFERENCES

1. B. S. Blanchard and E. E. Lowery, *Maintainability, Principles and Practices*, McGraw-Hill, New York, 1969.
2. C. E. Cunningham and W. Cox, *Applied Maintainability Engineering*, Wiley, New York, 1972.
3. Military Standardization Handbook MIL-HDBK-472, *Maintainability Prediction*, U.S. Department of Defense, 1966.
4. Military Standard MIL-STD-471, *Maintainability Verification/Demonstration/Evaluation*, U.S. Department of Defense, Revision A, 1973.

3

Availability

3.1 THE FUNDAMENTAL AVAILABILITY EQUATION

By this time, you should be convinced that MTBF (mean-time-between-failures) and MTTR (mean-time-to-repair) are useful design attribute measurement units. But you may find them somewhat awkward to relate to your product applications, particularly since you have to deal with each of them separately. Availability offers a single measure combining these two units into a single dimensionless unit that can be especially valuable for applications involving a product employed in substantial quantities by a given user. The *availability* of a product is calculated as follows:

$$\text{Availability} = \frac{\text{Up-time}}{\text{Up-time} + \text{down-time}}$$

where

up-time is the aggregate of all those time increments over a given service period during which the product is available for use (i.e., operating or in standby mode capable of operating);

down-time is the aggregate of all time increments over the same given service period during which the product is not available for use because it is inoperative, undergoing preventive or corrective maintenance.

If we restrict the time increments to those determined by its design, we can substitute MTBF and MTTR in the equation above, so that availability can be calculated from the following equation:

$$\text{Availability} = \frac{\text{MTBF}}{\text{MTBF} + \text{MTTR}}$$

3.2 HOW AVAILABILITY IS USED

The calculated availability value measures the proportion of the total service period that the product is actually available for use. Like MTBF or MTTR, availability is a statistical figure-of-merit more often applicable to all or a large part of a production lot, rather than to a single unit in that lot. The following example illustrates the use of the availability measure.

Consider an auto rental agency whose weekly demand level varies in generally predictable fashion—approximately 90 autos must be available on weekends, but less than 70 autos are needed during the regular work-week. Shop records indicate that only about 90% of the agency's auto fleet are in rentable condition at any one time because the other 10% are undergoing corrective (unscheduled) maintenance. The availability of this auto fleet is, therefore, 90%. Note that preventive maintenance need not affect the availability level at peak demand times because preventive maintenance tasks can be scheduled for the lower demand level periods during midweek.

It is reasonable to assume that the agency must operate a fleet of 100 autos in order to meet the weekend peak demand level of 90 autos. However, that 90-auto demand figure is based on previous experience, and is only a mean value, statistically speaking. A more sophisticated evaluation of the historic data would enable the agency to predict the variation in weekend demand to almost any desired confidence level, and the fleet could be sized to accommodate the maximum anticipated demand. However, it is safe to say that a cost analysis would demonstrate that, for maximum return on investment, the optimum fleet size would not be significantly greater than 100 units. The potential revenue lost from rental opportunities on a few busier-than-usual weekends would not offset the high acquisition and maintenance costs of additional autos. Therefore the mean demand figure is a satisfactory approximation to be used for sizing the fleet.

3.3 SOME CAVEATS

Now we know what availability is, and how an availability prediction can be used. But some warnings must be heeded in order to avoid possible pitfalls in applying this knowledge to real-world situations.

First, a stated requirement for an availability of 90%, in our auto rental agency example, could be met with an MTBF of 90 hours and an MTTR of 10 hours, or with an MTBF of only 45 minutes and an MTTR of 5 minutes. Obviously, an auto which has to be taken out of service every three-quarters of an hour of operation would irritate the agency's customers, even though the individual outage periods were very brief.

Second, unless the time span over which the availability requirement is to apply represents a substantial portion of the product's total service life, conflicting judgements are possible as to whether the requirement has been met. For example, an auto may satisfy a given availability specification over a full calendar year but not for the month within that year in which a major overhaul is performed.

Third, unconditional acceptance of MTBF as a measure of uptime will lead to an overly optimistic estimate of availability, since auto breakdowns will result from human errors and other unforeseeable external factors, such as stalling upon entering a flooded intersection. These are classified as non-chargeable failures, and are not counted when calculating MTBF's—only chargeable failures are counted; i.e., failures attributable to design deficiencies or manufacturing defects.

Fourth, unconditional acceptance of MTTR as a measure of downtime will also lead to an overly optimistic estimate of availability, since the MTTR does not include delays in repair while awaiting replacement parts, or time lost in daily shop routines (processing paperwork, working off backlogged repair jobs, coffee breaks, etc.). Furthermore, if the application demand level is constant, rather than cyclic (e.g., weekend peaks), preventive maintenance will have to be included in the aggregate down-time value.

After having made suitable allowances for the optimism built into MTBF and MTTR predictions, the designer is presumably left with a meaningful availability estimate—but what does it mean, really? Our example of the auto rental agency demonstrates that availability can be profitably applied to large groups of like items used in like ways. Availability can also be applied to a single unit intended for random-demand usage, where the product is expected to operate over a long service life made possible by a substantial number of repairs along the way. The familiar office paper copier fits this description. If a certain model has demonstrated an availability of 70%, you can expect that the machine to be unavailable 30% of the time. Suppose that the model can be redesigned to achieve an availability of 85%, which equates to an unavailability of only 15%, half that of the cheaper unit. But the redesign might be more expensive to produce, and would impose non-recurring development and retooling costs that would have to be amortized over the production run. Should your company undertake the changeover to the new model?

Assuming that your customers will accept your availability data and make their purchasing descisions on a strictly rational basis, your marketing people will have to evaluate the customers' copier needs on the basis of how much down-time costs them. From the customers' standpoint, the durations of individual down-times must be considered, as well as the aggregate down-time. In other words, the trade-off study will have to consider MTTR and M/max/ct, as well as availability, and established a scale of values expressing the worth, in dollars, of each minute of downtime eliminated by the design improvement (the 'disutility of downtime', in the language of the economist). A detailed trade-off study should employ the operations analysis methodology known as queuing theory. Fortunately, making the trade-off study will be the marketing department's headache, not yours. All you will have to do is furnish estimates of MTTR, M/max/ct, and availability.

One more thing remains to be said about availability. Like the reliability estimate from which it is in part derived, an availability estimate serves to predict the frequency of occurrence of down-times to be expected from units of a given design operating over an extended period of time. But the estimate provides no clue as to just what points in time units will be unavailable. Nor can we expect that every unit will exhibit the same level of availability over its service life. Availability is, after all, a statistically based estimating technique and must be used with that limitation in mind.

II
Maintenance Analysis

Maintenance activities, as we have already discovered, are of two kinds—preventive and corrective. Of the two, preventive maintenance offers the greater potential for the designer seeking to reduce maintenance costs, so preventive maintenance analysis is addressed first, in Chapter 4. But corrective maintenance possesses opportunities that should not be neglected, so corrective maintenance analysis procedures are given a chapter of their own, Chapter 5.

The climax of all this activity is reached in Chapter 6, where we learn how to transform the data generated by preventive maintenance analysis into phased-maintenance schedules, and how to use corrective maintenance data in planning factory support of fielded products, as well as in the preparation of repair and overhaul manuals.

4

Preventive Maintenance Analysis

4.1 INTRODUCING RELIABILITY CENTERED MAINTENANCE (RCM)

Until recently, the only basis for preparing maintenance instructions for new products was the instructions that had been used for previous models of similar designs. The writer of the instructions, who was in many instances also the designer of the product, was forced to rely on subjective judgement in selecting appropriate maintenance tasks and, particularly, in recommending how frequently each task should be performed. As a consequence, certain preventive maintenance tasks became standardized, remaining unchanged through many generations of model changes, even though they had been rendered inadequate or unnecessary by advances in product design technology. At the same time, new maintenance needs went unrecognized. Products—and their users—suffered because no technique was available for identifying the preventive maintenance actions specifically applicable to a new design.

The commercial airline industry was particularly concerned about the lack of a methodology for optimizing the scheduling of preventive maintenance because the profitability of their operations is very sensitive to the availability of aircraft during peak traffic periods, and because aircraft maintenance constitutes a substantial portion of their operating expenses. It is not surprising, therefore, that the now generally-accepted technique for optimizing preventive maintenance originated with the commercial airlines.

In Chapter 2 we defined preventive maintenance as the performance of those tasks that serve to defer or prevent an anticipated failure occurrence. *Reliability Centered Maintenance (RCM)* methodology is structured to implement the principle that no preventive maintenance task will be performed unless it can be justified, the criteria for justification being safety, availability, and cost-effectiveness in deferring or preventing a specific failure mode. (Implicit in these criteria is the premise that every mode in which the product might fail can be identified and analyzed to determine whether or not a corresponding preventive-maintenance task can be found.)

RCM is applied in the form of a two-part analysis of the product design. The first part of the analysis identifies potential failure modes, their sources, and their frequency of occurrence (failure rate). This part is actually a form of reliability evaluation known as *Failure Modes Analysis (FMA)*. The second part, the RCM analysis proper, addresses a logically sequenced set of questions to each failure mode. The answers to these questions determine the preventive maintenance tasks to be performed, and may disclose maintainability deficiencies that must be corrected.

It is apparent that the product's designer is the person best qualified to perform the RCM analysis. It is also true that the earlier an RCM analysis is performed, the more easily can its findings be accommodated in the product's design. One design area for which this rule is particularly apt is the incorporation of *built-in test circuitry (BITE)*, for detecting impending failures. For maximum operational effectiveness and least conflict with the basic configuration of the product, BITE requires early integration into the primary functional circuitry—certainly before the fabrication of any hardware, even prototype hardware. The optimal approach is to perform an initial analysis when the circuitry design first becomes available, followed by periodic analysis updates as the design is refined.

4.2 CONSTRUCTING THE ASSEMBLY HIERARCHY

We have already encountered the assembly hierarchy in Chapter 1, where it was described as the organization of system hardware elements into a structure that looks like the root system of a tree. The assembly hierarchy is developed by breaking the product down into successively lower lower-level assemblies until all piece-parts have been acounted for. An assembly hierarchy can be laid out in the form of an organization chart or as an indentured parts list. At the top level is the entire product, considered as a system that performs a designated mission or set of missions. The system is usually composed of a group of circuits, each of which performs specific functions that enable the system to perform its designated mission. Every circuit is made up of interconnected components, each of which is either a piece-part or an assembly that can be broken down into piece parts and/or subordinate assemblies. Those subassemblies can, in turn, be broken down into still lower-level piece-parts and/or assemblies. Ultimately, every line of descent in this root system must terminate in piece-parts.

Portions of an assembly hierarchy for an automobile are shown in Figures 4-1 and 4-2. (You have probably noted that we are using the automobile throughout this book to illustrate various points—the automobile provides the ideal example because everyone is familiar with its operation and because it exhibits virtually every conceivable kind of failure mode and maintenance need.)

4.3 ANALYZING COMPONENT MAINTAINABILITY

Each component included in the assembly hierarchy is analyzed, in turn, beginning at the system level and working downward. A separate RCM form is filled out for each component unless it is non-repairable; however, non-repairable components are included as failure sources in the RCM form for the next higher level assembly. Figure 4-3 illustrates an RCM form filled out for the radiator assembly of an automobile. Note that the specific component addressed by a given RCM form is designated as the ITEM UNDER ANALYSIS entry—in filling out the form, it is essential to remember that all questions must be answered in terms of the assembly hierarchy level on which the item under analysis is located. Thus, the NEXT HIGHER ASSEMBLY entry for the radiator assembly addressed by our sample form is the engine heat transfer subsystem. (The reason for using this formalized designation for the cooling system is to facilitate thinking about the hardware in terms of function rather than physical appearance.)

The FUNCTION(S) box in the heading of the RCM form should be filled out with particular care, since the function defined for the item under analysis will govern the analyst's reasoning in performing the analysis. In all but the simplest cases, the analyst may find himself going back to change a function entry as he

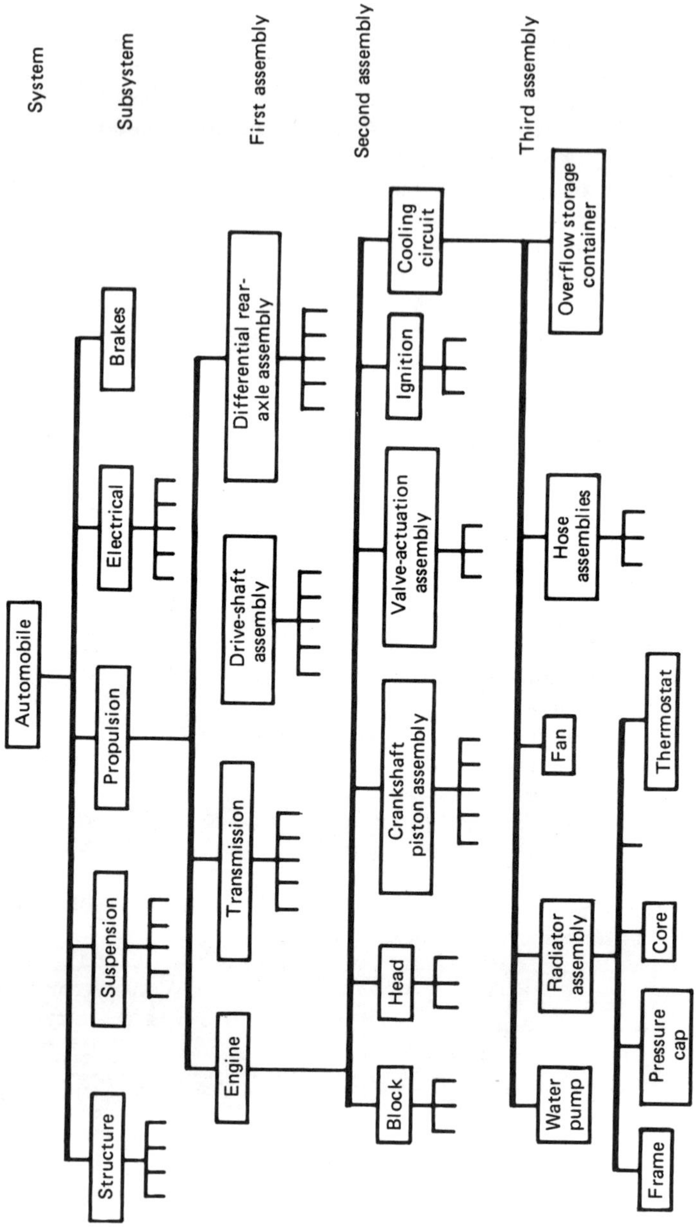

Figure 4-1 Portion of assembly hierarchy for an automobile engine cooling circuit.

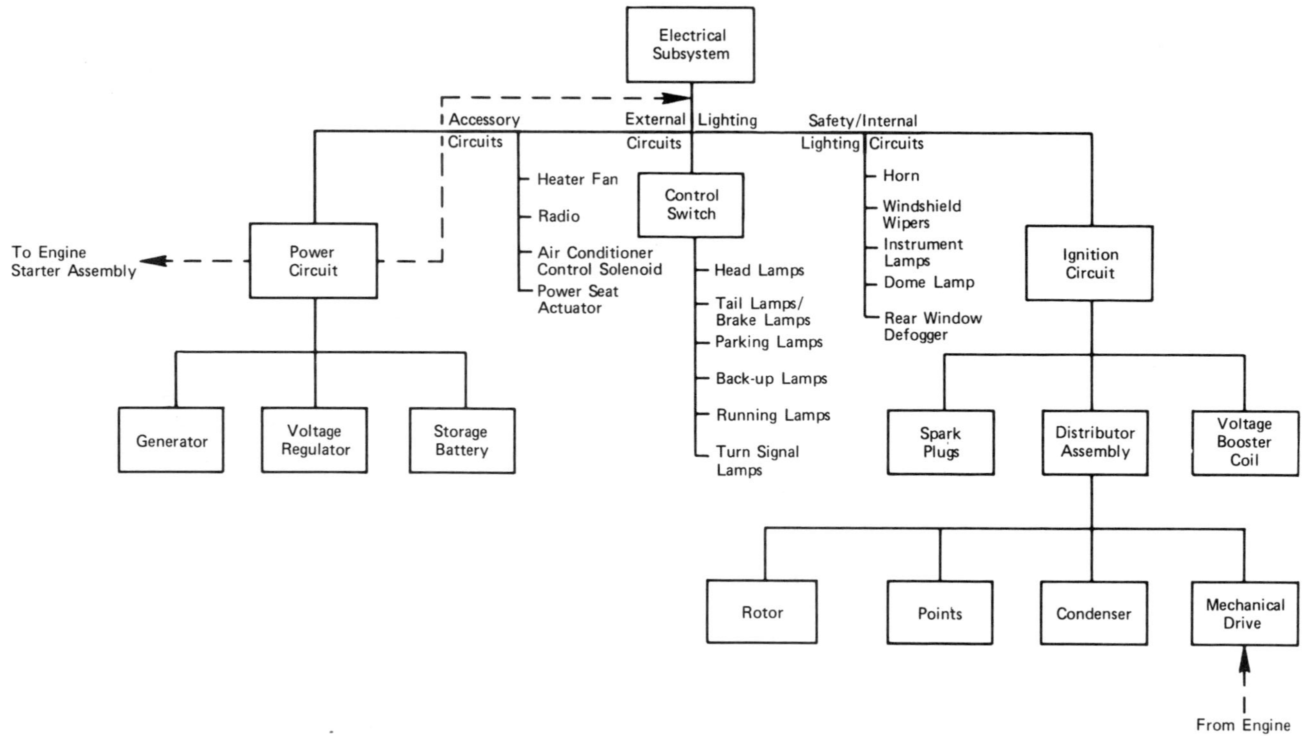

Figure 4-2 Portion of assembly hierarchy for an automobile electrical subsystem.

SHEET / OF /

ENTRY	NAME	PART NUMBER	FUNCTIONS PERFORMED BY ITEM UNDER ANALYSIS
ITEM UNDER ANALYSIS	Radiator assy.	903411	Transfers heat from coolant to ambient
NEXT HIGHER ASSEMBLY	Engine heat transfer S/S	900400	air
SUBSYSTEM	(same as NHA)		

FAILURE EVENT – 1.1 MODE		1.2 SOURCE CODE	1.2 SOURCE NAME	1.2 SOURCE NUMBER	1.3 RATE (FPMH)	1.4 LIFE LIMIT	1.5 EFFECT
1	Loss of coolant (leakage)	1.1	Radiator core	903423	1000		E-2
		1.2	Pressure cap	811-47R	500	5 yrs.	E-2
2	Restriction - blockage of flow	2.1	Radiator core	903423	240		E-2
3	Coolant temp. stays below spec. steady-state range	3.1	Thermostat	922173	3000		G-1

MODE	SERVICING – 2.1 TASK	2.2 PURPOSE	2.3 FREQUENCY
1	Check coolant level in overflow reservoir and radiator		once weekly/once monthly
2	Flush radiator per MP-1077	remove solid particles & gelled matter	once yearly
3	None		

MODE	MONITORING – 3.1 CONDITION	3.2 CRITERION	3.3 FREQUENCY
1	Coolant temperature (gage on dash)	> 180°F	5-minute intervals
2	(same as for Mode 1)	(same as for Mode 1)	(same as for Mode 1)
3	(same as for Mode 1)	< 155°F	5 minutes after each cold start

MODE	ASSESSMENT – 4.1 CONDITION	4.2 CRITERION	4.3 FREQUENCY
1	Puddling beneath vehicle	Presence of puddle	daily
2	Rate of draining during flushing	abnormally slow	once yearly (while flushing)
3	None		

Figure 4-3 Reliability centered maintenance (RCM) form filled out for the radiator assembly of an automobile.

progresses through the analysis of the item or even of a subordinate component several assembly levels lower. Because of the "ripple" effect that the function definition for a higher-level item has on the analysis effort downstream, the earlier a correction is made, the better. While this iterative correction process may be somewhat exasperating, it must be pursued faithfully if the analysis is to accomplish its purpose.

The next two sections of this chapter summarize the procedure for filling out the RCM form. Then we'll analyze the thought processes that led to the entries made on the forms shown in Figures 4-3, 4-5, and 4-6.

4.4 THE FAILURE MODES ANALYSIS PROCEDURE

Block 1 of the RCM form, titled FAILURE EVENT, actually constitutes what is generally called a Failure Modes Analysis (FMA). It is filled out as explained below—refer back to Section 1.3 for definitions of failure mode, failure source, and limited-life component:

Sub-block 1.1—(Failure) MODE

List in Sub-block 1.1 each mode in which the item under analysis could fail to perform its mission(s) or function(s), as applicable. Number these failure modes with sequentially assigned Arabic numerals in the left-hand column.

Sub-block 1.2—(Failure) SOURCE

For each failure mode listed in Sub-block 1.1, list in Sub-block 1.2 the name and part number of every next-lower-level component of the item under analysis that could, if faulty, cause that failure mode to occur. (Use a separate horizontal line for entering each source associated with a given failure mode.) Number these failure sources in the left column of Sub-block 1.2, using a coding sequence made up of the corresponding failure-mode number from Sub-block 1.1., followed by a period, and then by a sequentially assigned Arabic numeral. (This listing of the failure sources also serves to identify the items to be analyzed at the next lower assembly level.)

Sub-block 1.3—(Failure) RATE

For each failure source listed in Sub-block 1.2, enter the corresponding failure rate in Sub-block 1.3, expressed in failures per million hours (FPMH) relative to product operating hours. (If the failure rate data is expressed in some other unit (e.g., failures per million cycles), it must be converted to FPMH units for entry, except in the kind of situation where only a quick first-look evaluation is needed, or where quantitative failure-rate data is just not available for generic parts types being used in the design.

Where quantitative data is not available, relative measures based on user experience can provide useful information. These relative measures can be subjective judgements of comparable failure tendencies based on a scale of 1 to 10, or on a trinary scale (high, middle, or low), or even on a binary scale (high or low). Even binary-scale information will enable the analyst to pinpoint design elements requiring preventive and/or corrective maintenance.

Sub-block 1.4—LIFE LIMIT

For each failure source that is a limited-life item, enter in Sub-block 1.4 the corresponding life limit expressed in units that are directly comparable with the units in which the service life of the product is expressed.

Sub-block 1.5—(Failure) EFFECT

For each failure mode listed in Sub-block 1.1, enter in Sub-block 1.5 the appropriate code number to describe the criticality of the failure mode with regard to its effect on the mission of the system represented by the product, on the operator, and on the environment. The code number is constructed by selecting the appropriate alpha character from the left column of Table 4-1, followed by a hyphen and the appropriate numeral selected from the right column of Table 4-1.

The alpha character (left column) denotes the most probable consequence of an occurrence of the failure mode. (In ambiguous cases, it is usually advisable to select the most serious of the alternative possibilities). The Arabic numeral (right column) denotes the relative probability of occurrence of the selected consequence, and whether the consequence would follow immediately after the failure mode had occurred, or would be delayed long enough for the failure occurrence to be detected and appropriate corrective action taken. In other words, whether there is sufficient time for condition-monitoring to detect an impending failure and prevent the consequences from actually happening.

Consequences that occur immediately tend to be associated with catastrophic failures, i.e., occurrences where the physical failure mechanism is a sudden structural (or microstructural) yielding or transformation. Consequences whose effects progress slowly are degradative failures; i.e., occurrences where the physical failure mechanism is a gradual wearout or other deteriorative process.

Table 4.1 Failure Effects Classification Categories

Criticality level	Timing/probability of occurrence
A—Fatality or major injury	1— Immediate occurrence/certain
B—Minor injury	2— Impending occurrence unless condition corrected/certain
C—Major damage	3— Immediate occurrence/quite possible
D—Minor damage	4— Impending occurrence unless condition corrected/quite possible
E—Total loss of operating capability	
F—Loss of major operating mode capability, or substantial performance degradation	
G—Loss of minor operating mode capability, or significant performance	
H--Inconvenience	

Note in Table 4-1 that: (a) the ctiticality level categories are ranked in decreasing order of relative criticality; (b) the timing/probability-of-occurrence categories are ranked in decreasing order of certainty and immediacy; and (c) the category definitions listed are a generic compromise intended to accommodate many types of products, but can and should be modified by the analyst to achieve a closer fit for his particular situation.)

4.5 THE RCM ANALYSIS PROCEDURE

Filling out the rest of the form will complete the RCM analysis. Entries for the spaces in Blocks 2, 3, and 4 (and the life limits entered in Sub-block 1.4) will identify the preventive maintenance tasks that will serve to defer or prevent occurrence of each failure mode described in block 1. The analysis is performed by answering, for each failure mode listed in Sub-block 1.1, a set of ten logically interrelated questions that define the RCM flow path (Figure 4-4). Each question is answered either YES or NO. The

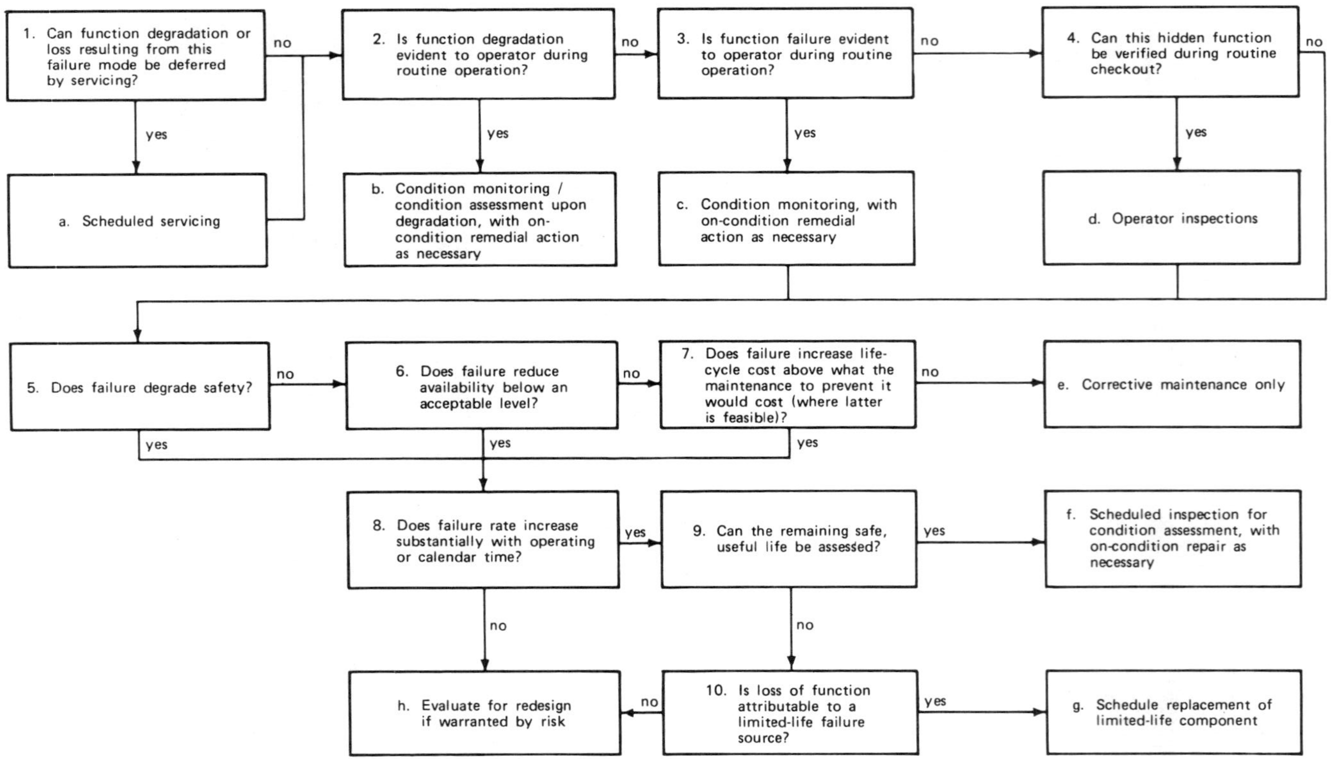

Figure 4.4 RCM flow diagram—questions and responses for identifying preventive maintenance tasks.

answer to each question leads to a preventive maintenance task or other conclusion for that failure mode—or to the next question. The answers to the questions determine the entries to be made on the RCM form.

Responses to each question should be made as described below. Although every failure mode of every component at every assembly hierarchy level must be evaluated, the task can be greatly simplified by lumping together all those items that can be treated in the same way (e.g., all structural elements with painted surfaces), and ignoring items that obviously require no preventive maintenance (e.g., structural elements that need no protective coating.)

Question 1: *"Can function degradation or loss resulting from this failure mode be deferred by servicing?"*

Every servicing or other preventive-maintenance task must serve to prevent or defer the loss or unacceptable degradation of a critical function performed by the item under analysis. For example, neglecting to check the coolant level in the radiator of an automobile could result in stoppage of and damage to the engine.

If the answer to this question is NO, proceed to Question 2. If the answer is YES, enter the appropriate task(s) in Sub-block 2.1 of the RCM form, if a suitable task(s) can be formulated. Enter the purpose of the task in Sub-block 2.2 only if it is not readily apparent for the Sub-block 2.1 entry. Enter the frequency with which the task should be performed in Sub-block 2.3. Then proceed to Question 2. *Note*: Be sure to identify the applicable failure mode in Blocks 2, 3, and 4 by entering the Failure Mode number from Sub-block 1.1 in the MODE Column.

Question 2: *"Is function degradation evident to operator during routine operation?"*

If the operator can sense degradation of the function resulting from the failure mode either directly (by observing a change in appearance, feel, smell, sound, etc.), or by means of an indicator (panel gage, caution or warning light, etc.), thereby detecting the onset of a failure before it happens, the answer to Question 2 is YES, and the failure mode is subject to condition monitoring. When an impending failure has been detected by condition monitoring, a condition assessment will then be performed by a repairman to determine whether on-condition repair is necessary.

If the answer to Question 2 is NO, proceed to Question 3.

If the answer to Question 2 is YES, enter the condition to be monitored in Sub-block 3.1, the frequency with which the condition is to be monitored in Sub-block 3.2, and the condition criteria that indicate an impending failure in Sub-block 3.3. This completes the analysis for the subject failure mode; proceed to the next failure mode.

Question 3: *"Is function failure evident to operator during routine operation?"*

If the answer is NO, go to Question 4. If the answer is YES enter the appropriate data in Section 3 of the RCM form, and then proceed to Question 4.

Question 4: *"Can the hidden function be verified during routine checkout?"*

If the answer to Question 3 was NO, the function related to the failure mode being analyzed is designated as a hidden function (see Item d in Section 2.6.1), and the item's ability to perform the hidden function must be verified by an inspection performed by the operator.

Typically, a hidden function is an emergency function that is not exercised during normal operation, such as fire-extinguishing. The operator should be able to verify certain types of hidden functions (e.g., smoke-detector alarm) by a routine test. In such cases the answer to Question 4 is YES—make the appropriate entries in Section 4 of the RCM form and proceed to Question 5. If the answer to Question 4 is NO, proceed to Question 5.

Question 5: *"Does failure degrade safety?"*

If the failure mode exposes the operator to risk of injury or the equipment to possible damage, the answer is YES. If YES, go to Question 8; if NO, go to Question 6.

Question 6: *"Does failure reduce availability below an acceptable level?"*

The product is usually unavailable for use whenever it is undergoing preventive or corrective maintenance. However, preventive maintenance (including replacement of limited-life parts) can often be scheduled for periods of inactivity such that availability is usually dependent only on corrective maintenance.

If preventive and corrective maintenance can be deferred to periods when the product is normally inactive, the answer to Question 6 is NO; proceed to Question 7.

If one or the other cannot be deferred, the answer depends on whether or not, the analyst's judgement, the reduction in availability is acceptable from the user's standpoint. If it is decided that the answer to Question is YES, proceed to Question 8.

Question 7: *"Does failure increase life-cycle cost above what the maintenance to prevent it would cost (where latter is feasible)?"*

In other words, would the consequences of permitting the failure mode to occur without hindrance cost the user more than it would cost him to perform preventive maintenance appropriate for reducing the frequency of occurrence? (The cost of failure, in this context, excludes the effects of this failure mode on other functions of the product, on performance of associated equipment, and on the success of the mission; these are broader issues, and should be considered in fundamental design cost/effectiveness studies.)

The answer to this question is of concern to the manufacturer of the product because it is important to the user. Life-cycle costing is a relatively new procurement consideration that offers fresh opportunities for an aggressive marketing organization, since minimal overall cost of ownership now becomes a selling point. Life-cycle costs are the aggregate costs of ownership and can, depending on the nature of the product, include such items as operating costs, housing costs, license fees, and ultimate disposal expence, in addition to purchase and maintenance costs. We are concerned here, of course, only with the maintenance portion of life-cycle costs. One cost element that must not be overlooked in pricing preventive maintenance is that of replacing limited-life components.

If the answer to Question 7 is YES, proceed to Question 8.

If the answer to Question 7 is NO, only corrective maintenance is to be performed; determination of the appropriate corrective maintenance task will be discussed in Chapter 5. This completes the analysis for the subject failure mode; proceed to the next failure mode.

Question 8: *"Does failure rate increase substantially with operating or calendar time?"*

"Failure rate," in this context, refers to the frequency with which the failure mode occurs as the aggregate of the failure rates of its component failure sources. As we know from the discussion of the bathtub curve in Section 1.5, the rates at which most parts fail tend to increase substantially after the part's operating life enters the wearout stage. If the failure rate characteristic of any of the failure sources associated with the failure mode does follow the bathtub curve, the answer to Question 8 is YES: proceed to Question 9.

If the answer to Question 8 is NO, the failure cannot be deferred or prevented by scheduled maintenance. Now, if the answer to Question 5 was NO, none of the failure modes of the item under analysis presents a risk beyond that of occasional inconvenient periods of unavailability, and the designer can decide whether or not to redesign strictly on the basis of additional cost versus enhanced product marketability.

But if the answer to Qestion 5 was YES, the question of risk assessment becomes more complex. If your automobile tends to stall and is difficult to restart, your risk can vary from mere inconvenience, while you are still in your own driveway, to possible fatalities if you are in the fast lane of the expressway. In the past, the designer had only his own judgement to guide him in assessing the degree of risk. Now, the designer has available the voluminous safety standards published by various professional societies and industry associations. Furthermore, his decisions are subject to review by various regulatory agencies and by the courts trying product liability suits.

If the designer decides that the risk is unacceptable, a redesign becomes mandatory. This redesign can take either of two approaches, depending on the nature of the product and its manner of application. For certain failure modes, an imminent failure can be detected before it actually occurs if the proper sensing device is installed and connected to a warning circuit (e.g., a strain gage mounted on a structural member and wired to set off an alarm horn when a predetermined threshold strain level is reached). This approach is acceptable if all that is needed is a warning given in sufficient time to prevent injury or damage.

If the warning approach is not suitable, the alternative is to reinforce the critical component or circuit with redundant units so that the effect of a critical unit failing is offset by the redundant component or circuit. The technique for incorporating redundancy is discussed in Chapter 7.

A redesign evaluation, in response to a NO answer for Question 8, completes the analysis for the subject failure mode; go to the next failure mode.

Question 9: "Can the remaining safe, useful life be assessed?"

Assessment in this case refers to condition assessment, as described in Item c of Section 2.6.1. A familiar example is measuring the thickness of a brake shoe to determine how many miles an automobile can be safely driven before the shoe must be replaced.

If the answer to this question is YES, enter the appropriate data in Section 4 of the RCM form.

Question 10: *"Is loss of function attributable to a limited-life failure source?"*

This question can be restated as: Is the safe, useful service life of any of the failure sources listed for the failure mode substantially less than the service life of the overall product? If the answer to Question 10 is YES, the action to be taken depends on which preceding question, 5, 6, or 7, called for a YES answer that led directly to Questions 8, 9, and 10.

If it was Question 5 (failure degrades safety), the life-ratings entered in Sub-block 1.2 for any such failure sources should be reviewed to verify that they incorporate an adequate margin for assuring that the failure sources are replaced before they fail normal preventive maintenance.

If it was Question 6 that drew a YES answer (failure reduces availability below an acceptable level), the life ratings for limited-life parts can be selected by means of a cost/effectiveness analysis to maintain availability at an acceptable level.

If the YES answer was addressed to Question 7 (failure boosts corrective maintenance costs above preventive maintenance costs), a cost/effectiveness analysis can be used to select limited-life ratings that minimize the combined maintenance costs.

4.6 USING THE RCM PROCEDURE

Using the RCM analysis procedure effectively requires that the analyst exercise considerable judgement. To illustrate the decision problems that may arise, let's take a look at the sample RCM analyses in Figures 4-3, 4-5, and 4-6.

The analysis of the radiator assembly (Figure 4-3) is reasonably straightforward, but even here we find a few surprises. Of the three failure modes listed, the first two simply prevent the radiator from performing its intended function of dissipating heat, but the third failure mode has the opposite result—it causes heat to be dissipated when such a process acts to delay the temperature from reaching the level at which the engine was designed to operate. Another factor that should be noted about these three failure modes is their relative timing and criticality of effect (Sub-block 1.5). The first two modes result in total loss of operating capability but their effects are designated as impending, whereas the third mode is said to have an immediate but minor effect.

The timing of the effect is determined from the condition-monitoring criterion (Sub-block 3.3), which can be said to establish a "chronological reference point." For the first two modes, the operator is alerted to the possibility of an impending failure as soon as the needle of the temperature gage begins to rise above the 180 degree F point. The analyst has assumed that the operator will be able to take appropriate corrective action before the engine overheats, and certainly before the pistons seize against the cylinder walls, causing major damage. (That is why the EFFECT code is given as "E-2", instead of "C-1").

For the third mode, however, the operator will not know that anything is wrong until the five-minute warm-up period has expired. Our analyst has interpreted this somewhat ambiguous situation (ambiguous with regard to the chronological reference point) to constitute an immediate effect, although another analyst may take a different point of view. Actually, it doesn't really matter which point of view is taken so long as a consistent viewpoint prevails throughout the analysis, and is reflected in the maintenance planning based on that analysis.

Further ambiguity can be found in the determination as to whether the effect of the thermostat failure on the operation of the automobile actually constitutes a significant performance degradation (Criticality Level "F"). The military services have a neat solution for this problem. All performance requirements are explicitly stated in a specification—if a maximum allowable engine warm-up time to reach 155 degrees F (Sub-block 3.3) is specified, failure to achieve that warm-up performance is classified as just that—a failure.

The entries in Blocks 2 and 3 are essentially self-explanatory. It might be noted that the analyst felt it useful to explain the purpose for flushing the radiator regularly, again a matter of personal judgement. It might also be noted that no servicing task or condition assessment is indicated for the third failure mode. It is not always feasible to devise a means of preventing or deferring a failure mode and, in this case, the consequences of a failure is not very critical. One more point—the entry calling for detection of pubbling beneath the vehicle (Sub-block 4.1 for Mode 1) illustrates the value of the pragmatic approach—no need to provide a special sensor when direct observation will do.

Well, so much for RCM analysis of the Radiator Assembly. With the Electrical Subsystem we find a more complex situation. The assembly hierarchy for the Electrical Subsystem (Figure 4.2) shows us why. You can see that the components of the Electrical Subsystem at the next lower level are not all of the same general kind—what we have here can be described as a "mixture of apples and oranges." The "apples" are two critical circuits, critical in the sense that a failure of either circuit would have the effect of terminating the principal system mission by causing the engine to quit. The "oranges" consist of a large number of relatively simple circuits grouped for convenience under the heading of Accessory Circuits, External Lighting Circuits, and Safety/Internal Lighting Circuits. The term, simple, in this context means that the components of the circuit are series-linked (see Chapter 7) and few in number, consisting in some cases of a single functional element and a circuit breaker or fuse.

Another set of complications for the analyst is introduced by certain interfaces among elements of the Electrical Subsystem and between elements of this subsystem and those of the Propulsion Subsystem. The Power Circuit supplies electrical power to all other circuits within the Electrical Subsystem (represented on Figure 4-2 by the broken line from the Power Circuit box to the line descending from the Electrical Subsystem box). The Power Circuit also supplies electrical power to the Starter Assembly. The analyst who prepared our assembly hierarchy chose to assign the Starter Assembly to the Engine Assembly, but this was a judgement on his part since he might just as well have classified it together with its interconnecting wiring, as a circuit within the Electrical Subsystem.

The rule in constructing assembly hierarchies that include circuits is to follow functional relationships rather than mechanical assembly relationships. That is why the spark plugs are grouped under the Electrical Subsystem, even though they are physically installed within the engine block. However, if the Starter Assembly is viewed from the functional standpoint, it does not clearly belong in either camp because its function is both electrical and mechanical. The broader line between the Electrical Subsystem and the Propulsion System falls within the Starter Assembly, which is not designed for disassembly. Therefore we shall have to add a codicil to our basic rule stating that, where two functionally different subsystems or circuits interface via a transducer (i.e., a device that serves to transform one kind of transmitted energy into another kind), the mechanical assembly relationship can be followed in constructing the assembly hierarchy. However, even this logic can fail us. Note that the distributor is driven by a mechanical drive that is powered, in turn, by the Engine Assembly. Our rule, with the codicil we have just formulated, would seem to indicate that the Mechanical Drive should be assigned to the Engine Assembly. But our analyst prefers to class the

Mechanical Drive under the Ignition Circuit on the purely subjective grounds that the function of the Drive is more directly related to the operation of the Distributor than to operation of the Engine Assembly. The ultimate rule for the analyst is, of course, to do it whatever the way enables him to do his job most effectively.

(If you are puzzled by the inclusion of such archaic devices as points and a condenser in the distributor instead of an electronic ignition module, or of a generator instead of an alternator, we must explain that our analyst is an old-timer who goes back to what he calls the "real cars" that we drove in the 1950s.)

Turning now to the RCM forms filled out for the Electrical Subsystem and its subordinate Power Circuit (Figures 4-5 and 4-6, respectively), we find a number of provocative entries that are worth considering individually:

1. Each of the failure modes listed on the RCM Analysis form for the Electrical Subsystem (Figure 4-5; Sub-block 1-1) comprises a simple loss of capability to perform one of the functions of its next-lower-level components. (For our illustration, each failure mode listed encompasses an entire group of functions; in an actual analysis, each individual function would have to be addressed separately.) Further details are not necessary at this high assembly hierarchy level—ambiguity is minimized by leaving them to the specific level at which the function is performed—in this case, to the next-lower-level circuits listed as failure sources. The "N/L" entries in Blocks 2.1, 3.1, and 4.1 of Figure 4-5 mean "next (lower) level," and indicate that specific information entries are deferred to the RCM forms for failure modes of items at the next lower assembly hierarchy. (Figure 4-6 applies to one of the next-lower-level items in our example.)

2. In the first failure mode, inclusion of the phrase, "on demand," denotes that the function lost because of the failure mode is required only at certain times in the operating cycle, rather than continuously throughout the cycle. This distinction can affect selection of the Arabic-numeral part of the code entered in Sub-block 1.3 that indicates the timing of the consequences relative to the failure mode occurrence. For a product like an automobile, where the operating cycle is a sequence of starting, operation, and shutdown time segments, a failure is considered to occur at some unpredictable moment during the operating time segment; for an on-demand function, however, the failure is treated as having occurred at the moment the function was demanded, although it may actually have occurred much earlier. Hence, the timing for on-demand failures is entered as an immediate occurrence. This

is an arbitrary rule of convenience, formulated to maintain consistency among effects entered for the different failure modes.

In our Electrical Subsystem example, the criticality-level part of the EFFECTS code entries in Sub-block 1.5 (i.e., the alpha character denoting consequences) are each generalized to cover an entire group of related circuits (exterior lighting, interior lighting/safety, comfort/convenience), and some of these entries may very well be contradicted at the next lower level, where effects of failure modes on system missions are addressed separately for each circuit. The effect of a failure mode at the lower level may completely eliminate the ability of a given circuit to perform its function, but the loss of that function need not necessarily have a serious impact on the principal function(s) performed by the higher-level subsystem.

At this point it may be well to recall that the functions of the Electrical Subsystem are by no means equally critical—they cascade in relative importance from the prime function (that of making operation of the automobile possible) all the way down to the convenience function of lighting the driver's cigarette. But in the eyes of the automobile dealer, each comfort and convenience feature can be sold individually and, therefore, constitutes a distinct, if optional, system mission.

Examining the specific entries in Sub-block 1.5 we find that, for the first failure mode, the effect of the failure mode is loss of major operating capability, occurring immediately and with certainty. Without sufficient electrical power, the engine cannot be started; without suitable spark ignition, the engine cannot be operated. Our analyst does not consider these types of failures to result in total loss of operating capability because they are separated, in his evaluation, from loss of other function capabilities which he addresses in the remaining three failure modes listed. He considers all auto driving functions as major operating modes—all other functions are minor operating modes. In support of his decisions he points out that the storage battery could be too weak to start the engine but still be able to power emergency lights and a CB radio that the driver uses to summon help.

Determining the effects of the remaining three failure modes requires even more subjective judgement. Loss of an exterior lighting function is interpreted as preventing night driving, which is a major operating mode, and the effect is immediate and certain: Loss of an interior lighting or safety function capability presents a more difficult judgement situation. If, in keeping with good design practice, we consider the worst possible situation here, it is clearly the loss of a safety function, probably that of windshield

SHEET 1 OF 2

ENTRY	NAME	PART NUMBER	FUNCTIONS PERFORMED BY ITEM UNDER ANALYSIS
ITEM UNDER ANALYSIS	Electrical S/S		a. Deliver electrical power to Engine Starter Assy.
NEXT HIGHER ASSEMBLY	Automobile		b. Deliver spark ignition to engine cylinders via plugs
SUBSYSTEM	(same as IUA)		c. Perform lighting, safety, convenience, etc., functions

	FAILURE EVENT – 1.1 MODE	1.2 SOURCE CODE	1.2 SOURCE NAME	1.2 SOURCE NUMBER	1.3 RATE (FPMH)	1.4 LIFE LIMIT	1.5 EFFECT
1	Unable to deliver sufficient power for starting	1.1	Power circuit		9320	N/L	F-1
2	Unable to deliver sufficient voltage to spark plugs	2.1	Ignition circuit		7240	N/L	F-1
3	Unable to deliver sufficient power to headlights	3.1	Headlight circuit		9320	N/L	F-1
4	Unable to deliver sufficient power to safety light	4.1	(specific circuit)		12,860	N/L	F-3

MODE	SERVICING – 2.1 TASK	2.2 PURPOSE	2.3 FREQUENCY
		N/L	

MODE	MONITORING – 3.1 CONDITION	3.2 CRITERION	3.3 FREQUENCY
		N/L	

MODE	ASSESSMENT – 4.1 CONDITION	4.2 CRITERION	4.3 FREQUENCY
		N/L	

SHEET 2 OF 2

ENTRY	NAME	PART NUMBER	FUNCTIONS PERFORMED BY ITEM UNDER ANALYSIS
ITEM UNDER ANALYSIS			
NEXT HIGHER ASSEMBLY			
SUBSYSTEM			

FAILURE EVENT – 1.1 MODE		1.2 SOURCE			1.3 RATE (FPMH)	1.4 LIFE LIMIT	1.5 EFFECT
		CODE	NAME	NUMBER			
5	Unable to deliver sufficient power	5.1	(specific circuit)			N/L	G-1
	to specific convenience or						
	entertainment accessory						

MODE	SERVICING – 2.1 TASK	2.2 PURPOSE	2.3 FREQUENCY

MODE	MONITORING – 3.1 CONDITION	3.2 CRITERION	3.3 FREQUENCY

MODE	ASSESSMENT – 4.1 CONDITION	4.2 CRITERION	4.3 FREQUENCY

Figure 4.5 Sample completed RCM form for automobile electrical subsystem.

SHEET 1 OF 2

ENTRY	NAME	PART NUMBER	FUNCTIONS PERFORMED BY ITEM UNDER ANALYSIS
ITEM UNDER ANALYSIS	Power circuit		Generates, stores, and delivers electrical power for starting engine and operating all other circuits within Electrical S/S
NEXT HIGHER ASSEMBLY	Electrical S/S		
SUBSYSTEM	(Same as NHA)		

	FAILURE EVENT – 1.1 MODE	1.2 SOURCE CODE	1.2 SOURCE NAME	1.2 SOURCE NUMBER	1.3 RATE (FPMH)	1.4 LIFE LIMIT	1.5 EFFECT
1	Unable to store energy sufficient to start engine	1.1	Storage battery		800	4 yrs.	F-2
2	Unable to fully recharge storage battery during operation after engine start	2.1	Generator		1350		F-2
		2.2	Voltage reg. (failed open)		620		F-2

MODE	SERVICING – 2.1 TASK	2.2 PURPOSE	2.3 FREQUENCY
1	Replenish electrolyte in storage battery cells as necessary		Once monthly
2	None		

MODE	MONITORING – 3.1 CONDITION	3.2 CRITERION	3.3 FREQUENCY
1	Generator output current	Immediately following engine start or heavy battery drain, ammeter drops from 10-15 amps to 0.3-0.8 amp within 5 minutes; remains at 0.3-0.8 amp thereafter	Continually throughout battery recharge period; 5-minute intervals thereafter
2	None		

MODE	ASSESSMENT – 4.1 CONDITION	4.2 CRITERION	4.3 FREQUENCY
1	Acid content of battery electrolyte	Hydrometer reading 1.3 ± 0.05	6-month intervals or when weak battery indicated in starting

SHEET 2 OF 2

ENTRY	NAME	PART NUMBER	FUNCTIONS PERFORMED BY ITEM UNDER ANALYSIS
ITEM UNDER ANALYSIS			
NEXT HIGHER ASSEMBLY			
SUBSYSTEM			

FAILURE EVENT – 1.1 MODE		1.2 SOURCE CODE	1.2 SOURCE NAME	1.2 SOURCE NUMBER	1.3 RATE (FPMH)	1.4 LIFE LIMIT	1.5 EFFECT
3	Unable to deliver power on demand for engine starting	4.1	Interconnection wiring	—	450	3 yrs.	F-1
4	Unable to deliver power on demand for other engine circuits	5.1	Interconnection wiring (not same wiring as 4.1)	—	500	3 yrs.	F-1

MODE	SERVICING – 2.1 TASK	2.2 PURPOSE	2.3 FREQUENCY
3	None		
4	None		

MODE	MONITORING – 3.1 CONDITION	3.2 CRITERION	3.3 FREQUENCY
3	None		
4	None		

MODE	ASSESSMENT – 4.1 CONDITION	4.2 CRITERION	4.3 FREQUENCY
3	None		
4	None		

Figure 4.6 Sample completed RCM form for automobile electrical power circuit.

wiper operation. In a rainstorm the auto could not be driven, which is a major loss of operating capability. Although most driving is done when it is not raining, the windshield-wiping capability is an on-demand function and, therefore, considered to occur immediately. But, since the conditions under which the windshield wipers are turned on vary considerably (rain intensity, prevailing light, dirt on windshield, etc.), the driver may still be able to drive without the wipers. Therefore, the "timing/probability" of the consequences is considered to be "immediate occurrence/quite possible."

3. No entries have been made in Blocks 2, 3, or 4 for any of the five Electrical Subsystem failure modes listed because these details should be addressed, as indicated previously, at the lowest relevant level.

4. A single function is entered on the RCM analysis form for the Power Circuit (Figure 4-6), but it is divided into three subfunctions—"generate," "store," and "deliver." Failure to perfcrm any one of these three subfunctions results immediately or eventually in failure of the Power Circuit to perform its function and, consequently, in the termination of all Automobile System missions. The value of stating the function in such a fashion is illustrated by the neat correlation between the function statement and failure mode entries—each of the latter is expressed in terms of loss of ability to perform one of the three subfunctions.

5. The failure modes listed for the Power Circuit parallel those listed for the Electrical Subsystem, but the timing of the first Power Circuit failure mode differs from that of its Electrical Subsystem counterpart in that it does not constitute the loss of an on-demand function since energy storage occurs only after the engine has been started; therefore, the timing of the consequence (inability to start) is not immediate but only impending—the consequence will not occur until the next time the driver attempts to start the engine.

The second failure mode requires an examination of the role played by the Voltage Regulator as a failure source. Although we have treated the Voltage Regulator as an independent component of the Power Circuit up to this point, since that is the way it must be shown on a reliability block-diagram model, we here recognize that the Voltage Regulator operates in conjuction with the Generator to charge the Storage Battery—that is its sole function. If the Voltage Regulator fails in the open-circuit mode, the Storage Battery will not remain charged even though the Generator is functioning properly. And if the Voltage Regulator fails in the closed-circuit mode, Interconnecting Wiring will be burned up, resulting in the fourth failure mode entered for the Power Circuit.

Consequential-damage situations can occur with many circuit designs, frequently making the task of determining the initial cause of of failure quite difficult, although not in this particular case. If an investigator finds that both the Voltage Regulator and associated Interconnecting Wiring are burned out, he is safe in assuming that it was the Voltage Regulator that first failed, thereby acting to short the Storage Battery terminals across the Wiring.

The third failure mode is unique in our analysis in that it is the only failure mode designated as resulting in total loss of operating capability. As previously noted, the Storage Battery can temporarily power the Engine Starter Assembly and all of the other Electrical Subsystem circuits, including the Ignition Circuit, even if the Generator fails.

The fourth and fifth failure modes are similar except that the Interconnecting Wiring entered as the failure source is different wiring in each case.

6. The designation of the Storage Battery as a limited-life component with a service duration of four years was made by the analyst as a somewhat arbitrary judgement, since an evaluation of the bathtub curve for a given model of storage battery would probably not show a marked upturn point indicating the onset of wearout. The duration of useful life for a storage battery is very sensitive to operating and environment stress levels, which can vary drastically from one user to the next. The manufacturer's intent here is to pass the limited-life information along to the customer in hopes that he will replace the battery before it fails, since a failure of any kind may be associated in the customer's mind with a defective product.

7. Only one of these failure modes has an associated servicing task—like most servicing tasks it involves checking, periodically, the quantity remaining of a necessary consumable, and replenishing as necessary to restore the quantity to a predetermined level.

8. Two kinds of condition-monitoring tasks are performed for the Power Circuit. One of the conditions monitored, that of Generator Output Current, nominally measures the performance of one of the Power Circuit's components rather than that of the Power Circuit as an entity. Actually, however, it enables the operator to detect degradation and impending failure modes for all of the components. If the Storage Battery will not charge or the Voltage Regulator fails in the open-circuit mode, the ammeter needle drops to zero; if the Voltage Regulator fails in the closed-circuit mode, the ammeter needle will go to a high negative value. By the way, the "continual" entry for frequency does not mean that the operator must not take his eyes off the ammeter, it merely indicates that he should scan the ammeter as frequently as is convenient.

Unlike the condition-monitoring entry for the first and second failure modes, the entry for the third mode does not involve the use of an explicit monitoring device but relies instead upon the operator's own equipment, which can see, hear, feel, smell and even taste departures from normal operation. We designers frequently fail to exploit the full capabilities of this fantastically sensitive and flexible condition-monitoring system. The "on demand" entry for frequency means that monitoring can only be performed when operation is demanded of a given circuit by the operator.

9. The Power Circuit is the lowest level along its line of assembly hierarchy descent that requires an RCM analysis, since all of the components at the next lower level are piece-parts and have no component failure sources.

4.7 APPLYING THE RCM ANALYSIS RESULTS

All of the information required for planning preventive maintenance of the product can be obtained from the completed set of RCM forms. However, one more step must be taken to arrive at an optimized preventive maintenance schedule for inclusion in the operating and maintenance manual supplied with the product—this step is discussed in Chapter 6.

5
Corrective Maintenance

5.1 THE NEED FOR CORRECTIVE MAINTENANCE

It might be possible, in theory, to eliminate any need for corrective maintenance by designing products that do not suffer from failures, employing the appropriate combination of high-reliability components, built-in test provisions (BITE), and preventive maintenance. But garage mechanics and other technicians who make their living by repairing the products you design have never had to worry about being designed out of a job. When product reliability is pushed beyond the level prevailing in general industrial practice, further incremental reliability improvements will increase production costs exponentially such that the marketing edge offered by promising the customer large savings in repair expenses is soon outweighed.

Perhaps the only exceptions to this rule are items that are unusually costly or difficult to repair, such as earth-orbit satellites which must operate without maintenance of any kind throughout their entire service lives—or at least until NASA's Space Shuttle achieves routine operational status. Since we cannot eliminate corrective maintenance, what we must devise is a method for generating the information needed for corrective maintenance planning, just as we devised a method in Chapter 4 for generating the information needed for preventive maintenance planning.

5.2 THE CMA ANALYSIS PROCEDURE

Our *Corrective Maintenance Analysis Procedure (CMA)* must identify the specific corrective maintenance tasks to be performed for each failure mode. It must also determine the disposition to be made of faulty items replaced in the course of correcting the failure—are they to be repaired or discarded?

The CMA procedure parallels the logically-sequenced-question approach taken in Chapter 4 for RCM preventive maintenance analysis, but with an important difference: the RCM questions were addressed, in turn, to every failure mode of a function performed by the item under analysis, but the CMA questions (see Figure 5-1) are addressed, in turn, to every failure source for every failure mode.

The information developed in answering the corrective-maintenance question sequence is entered on the CMA form shown in Figure 5-2. (A separate form is completed for each hardware item at each assembly hierarchy level down to but not including non-repairable items such as piece-parts.) The CMA form should be treated as an extension of the RCM form shown in Figure 4-2, and the block numbering used on that form continued in sequence on the CMA form. Figure 5-2 illustrates the corrective maintenance analysis performed on the auto radiator assembly used as an example in Chapter 4.

After entering the heading information, begin the analysis by listing the failure source components of the item under analysis in the FAILURE SOURCES column, identifying each by means of the corresponding code taken from the left-hand column of Sub-block 1.2 on the RCM form. The analysis is performed by subjecting each failure source, in turn, to the sequence of questions outlined in Figure 5-1, and documenting the answers to those questions in Blocks 5 and 6 of the CMA form, as described further on. The abbreviations in the column heading for Blocks 5 and 6 are explained below; the corrective maintenance terms used are defined in Section 2.6.2.

Block 5: Corrective Action

- AAC Adjust, align, or calibrate the failure source (no rework or part replacement required).
- RIP Repair the failure source in-place by physically reworking it. Although this category of corrective action is intended to apply to repair of the failure source while it remains in its installed location, the category must also include repair of sources removed from their installed

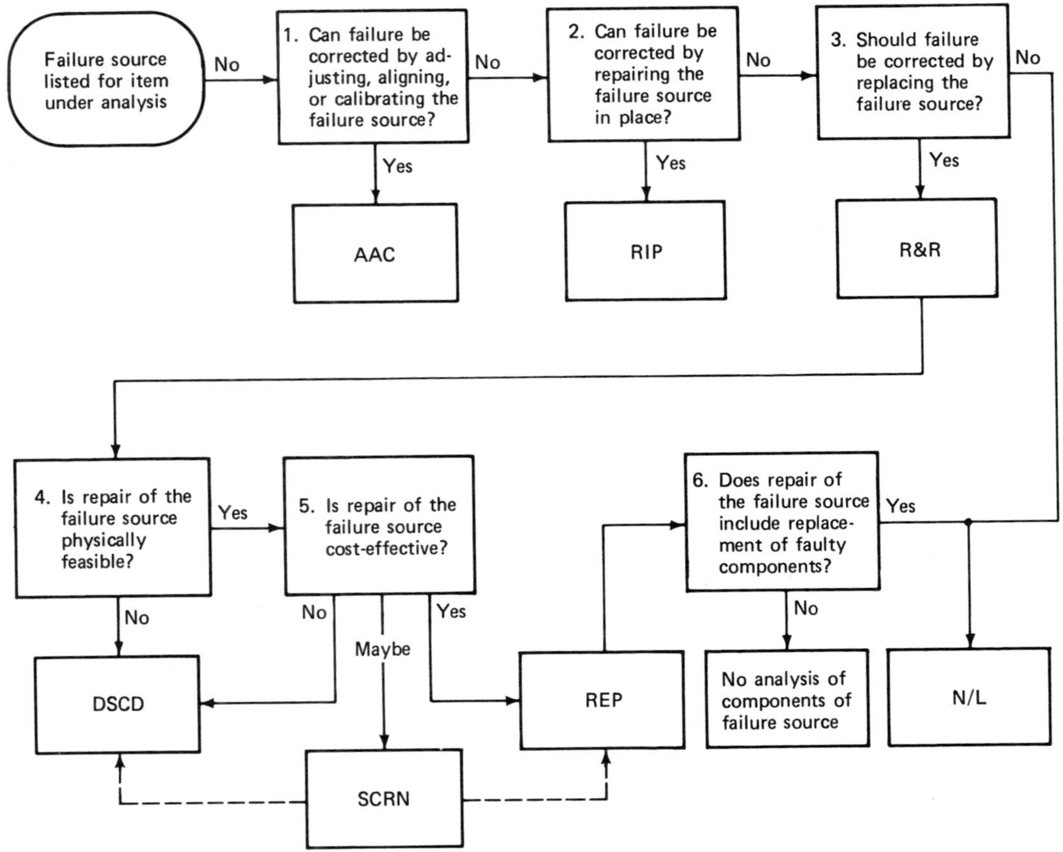

Figure 5.1 CMA Flow Diagram – Questions and Responses for Identifying Corrective Maintenance Tasks

SHEET OF

	NAME	PART NUMBER
ITEM UNDER ANALYSIS	Radiator assembly	903411
NEXT HIGHER ASSEMBLY	Engine heat transfer s/s	900400
SUBSYSTEM	(same as above)	

FAILURE SOURCE	5. CORRECTIVE ACTION				6. DISPOSITION OF FAILURE SOURCE				REMARKS
	AAC	RIP	R&R	N/A	DSCD	SCRN	REP	N/L	
1.1			X		X	X			
1.2			X		X				
2.1		X					X		RIP or R&R depending on extent of damage; REP if R&R
3.1			X		X				
Y.1	X			X					
Z.1								X	

Figure 5.2 Corrective Maintenance Analysis (CMA) Form Filled for Radiator Assembly of an Automobile

position in order to facilitate the rework, provided that all of the repair task can be accomplished immediately adjacent to the installed location (e.g., removing spark plugs from an automobile engine for cleaning and regapping).

N/A Not applicable (the appropriate corrective maintenance action is accomplished by AAC or RIP of the failure source, or is performed at an assembly hierarchy level below that of the failure source; in either case, repair of the failure source away from its installed location is not applicable).

R&R Remove the failure source from the item under analysis and replace it with a properly operating equivalent unit.

Block 6. Disposition of Failure Source

REP Repair the failure source.

DSCD Discard the failure source.

SCRN Screen (i.e., examine) the failure source to determine whether repair or discard is more cost-effective.

N/L Next level (an "X" in this column indicates that the failure source is to be analyzed; i.e., to be treated as one of the items under analysis when the analysis reaches the next lower assembly hierarchy level).

5.3 FILLING OUT THE CMA ANALYSIS FORM

Responses to each question asked in Figure 5-1 should be made as explained below. The first three questions deal with corrective maintenance performed on the faulty failure source; the following three questions address the disposition to be made of the faulty failure source.

Question 1. "Can failure be corrected by adjusting, aligning, or calibrating the failure source?"

The answer to Question 1 will be YES only if the failure can be corrected with no physical rework or replacement of any of its components.

If the answer is YES, enter an "X" in the AAC Column of Block 5 of the CMA form (shown in Figure 5-2). That ends the analysis for this failure source—go on to the next failure source listed.

If the answer is NO, go on the Question 2.

Question 2. "Can failure be corrected by repairing the failure source in-place?

The answer to Question 2 will be YES if the failure source can be repaired in-place by physical rework, with no replacement of any of its components.

If the answer is YES, enter an "X" in the RIP Column of Block 5. That ends the analysis for this failure source; go on to the next failure source listed.

If the answer is NO, go on to Question 3.

Question 3. "Should failure be corrected by replacing the failure source?"

The NO answers to Questions 1 and 2 tell us that correction of the failure cannot be accomplished by adjusting or repairing the failure source in-place. Therefore, it must be assumed that the failure can be corrected only by replacing the ultimate failure source; i.e., the failure source that actually suffered physical failure. The ultimate failure source may be either the failure source currently being analyzed or one of its components. The answer to Question 3 will be YES if it is not feasible to replace the ultimate failure source without replacing the failure source currently being analyzed.

If the answer is YES, enter an "X" in the R&R Column of Block 5 and go on the Question 4.

If it is feasible to replace the ultimate failure source by replacing a component at any assembly level below the level of the source being analyzed, the answer will be NO. In that case, enter an "X" in the N/A Column of Block 5. That ends the analysis for this failure source, but analyses will have to be performed of failure sources at the next lower assembly level.

Question 4. "Is repair of the failure source physically feasible?"

Certain failure sources will always be rendered unrepairable by a failure. Take, for example, the common incandescent light bulb—its expected failure mode is burn-out of the filament—repair not feasible. Another failure mode is shattering of the glass outer envelope—repair even less feasible. Catastrophic failures are generally non-repairable. On the other hand, wearout failures can be designed to be usually repairable.

If the repair of the failure source is physically feasible, enter an "X" in the REP Column of Block 6 and go on to Question 5.

If repair of the failure source is not physically feasible, enter an "X" in the DSCD Column of Block 6. That ends the analysis for this failure source–go on to the next failure source listed.

Question 5. "Is repair of the failure source cost-effective?"

There are three possible answers to this question–YES, NO, and MAYBE. The answer will be YES or NO in those cases where the consequences of the failure mode are consistent and unambiguous. But a more complex situation arises with failure sources in which the extent of damage resulting from a failure cannot be precisely limited by design–with such failure sources, the feasibility of repair must be assessed separately for each instance. This case-by-case assessment can only be made on the basis of examining the hardware after the failure has occurred; this is what is meant by *screening*.

If the answer to Question 5 is NO, the failure source should be discarded; enter an "X" in the DSCD Column of Block 6. That ends the analysis for this failure source–go on to the next failure source listed.

If the answer to Question 5 is YES, the failure source should be repaired; enter an "X" in the REP Column and go on to Question 6.

If the answer to Question 5 is MAYBE, the decision as to whether to discard or repair the failure source can be made only after examining it; enter an "X" in the SCRN Column. That ends the analysis for this failure source–go on to the next failure source listed.

Question 6. "Does repair of the failure source include replacement of faulty components?"

If the answer is NO, it will not be necessary to include the components of the failure source in the analysis of this failure mode at lower assembly hierarchy levels. That ends the analysis for this failure source–go on to the next failure source.

If the answer is YES, enter an "X" in the N/L Column of Block 6, indicating that the analysis will include components of the failure source at the next lower assembly hierarchy level. That ends the analysis for this failure source–go on to the next failure source listed and complete the analysis for all failure sources at this level before moving to the next level down.

5.4 PERFORMING THE CMA ANALYSIS

Let's check out the CMA procedure by analyzing the auto radiator assembly for which we did an RCM analysis in Chapter 4. The RCM analysis form in Figure 4-3 lists three failure modes for the radiator assembly, which are referred to by their corresponding failure mode numbers on the CMA form shown in Figure 5-2.

The first failure mode is loss of coolant by leakage. Two failure sources are listed for this failure mode: 1.1 (Radiator core) and 1.2 (Pressure cap). Taking 1.1 first, we decide that the answers to Questions 1 and 2 are both NO, because the failure cannot be corrected by adjustment or repair in-place. Since a radiator leak cannot be repaired by replacing a subordinate component, the answer to Question 3 must also be NO. Therefore, an "X" is entered in the R&R Column.

Questions 4 and 5 call for value judgements. Repair of the radiator is generally feasible and, more often than not, cost-effective. The entries on the CMA Form in Figure 5-2 show the answers to Questions 4 and 5 to be YES and MAYBE, respectively; hence the "X" in the SCRN Column. Your judgement, however, may lean towards a YES answer on Question 6, reasoning from your experience that it is actually more cost-effective to treat all radiator leakage cases as repairable than to call for a detailed screening in every case. You are the analyst and your own judgement must prevail, the only absolute rule that should govern your judgements is the rule of consistency—all similar cases must be decided the same way or you will lose the confidence of the people who have to apply your maintenance decisions.

Failure Source 1.2 (pressure cap) follows the same logic trail as Failure Source 1.1 through the first three questions, but the answer to Question 4 is NO; hence the "X" in the DSCD Column.

The second failure mode has only one failure source: 2.1 (Radiator core). The method to be selected for correcting a blockage in the radiator core depends on the extent of the blockage. A minor blockage can be cleaned out without removing the radiator from under the hood of the automobile, but a major blockage cannot. Therefore, an "X" is entered in the RIP Column of Block 5, and another "X" in the REP Column of Block 6.

This situation illustrates a fundamental hazard in following logic diagrams such as our CMA diagram in Figure 5-1 blindly. The diagram is predicated on making either/or decisions. For most cases, the either/or approach is fine. But here we have a situation where both choices can be correct—the proper answer to Question 2 is YES AND NO. Our logic diagram could be expanded to accommodate such anomalous situations, but if we did that we might have to expand it in other places to accommodate other anomalous situations. So, in designing the CMA Flow Diagram, we favored simplicity over completeness, reasoning that it is more cost-effective (in terms of the cost of your analysis labor) to limit coverage of the diagram to the usual situation rather than trying to accommodate all possibilities. This means, of course, that you have to use our diagram with discretion (or construct your own more-complicated version).

The entries for Failure Source 3.1 (thermostat) are the same as for Failure Source 1.2 (pressure cap), since it is generally not feasible to repair a faulty thermostat.

The foregoing example did not illustrate all of the possible column entries, so let's clean those up on an individual basis. Persistent stalling of an automobile engine can be considered to be a failure of the Throttle Linkage Assembly; let's designate this failure source as Y.1 on the CMA form in Figure 5-2. If the stalling can be eliminated by adjusting the Throttle to increase the idling speed, "X'es" are entered in the AAC and N/A Columns of Block 5, the latter "X" indicating that no item has been replaced as faulty and, therefore, no entry is needed in Block 6.

As for the N/L Column of Block 6, let us consider Failure Source 3.1 (corresponding circuit) in Figure 4-5 (Electrical Subsystem). Corrective maintenance is performed not on a circuit, which is merely a concept, but on specific hardware components of a circuit. In such cases, an "X" is entered in the N/L Column, thereby extending the analysis to the failure-source components which make up the next-lower-hierarchy-level circuits comprising the Electrical Subsystem. (To avoid confusion with the other Failure Source 3.1 already shown in Figure 5.2, we'll redesignate this failure source as Z.1 on our sample CMA form.)

5.5 APPLYING THE CMA ANALYSIS RESULTS

The completed CMA forms identify the corrective maintenance tasks for which the manufacturer will have to provide support in the form of overhaul and repair manuals; spare parts; special materials, tools, test equipment, and technician training programs, as necessary. We'll discuss these in greater detail in Chapter 6.

6
Maintenance Planning

6.1 THE CASE FOR MAINTENANCE PLANNING

Formal maintenance planning probably originated as a part of the *logistics* planning performed by military organizations to achieve the timely provision of material, personnel, and facilities required for wartime operations. (These activities are now designated as *Integrated Logistic Support (ILS)* by the U.S. Department of Defense.) Until recently, however, manufacturers of industrial, commercial or consumer hardware have generally not felt it necessary to rethink their maintenance strategy when introducing each new product, relying instead on traditionally evolving methods and procedures. Virtually all manufacturers of household appliances support the maintenance of their products in the same way; so do all automobile manufacturers, most office equipment manufacturers, hydraulic press manufacturers, and so on. But, as we have seen when discussing the need for RCM in Chapter 4, depending on tradition can be unnecessarily costly to a manufacturer's customers—costly in terms of both maintenance expense and product availability when needed. To assure that a given product is supported by the most economical and effective combination of maintenance procedures and resource provision feasible, a formal approach is needed.

6.2 THE SCOPE OF MAINTENANCE PLANNING

Maintenance planning should begin with the data compiled by the preventive and corrective maintenance analyses described in Chapters 4 and 5, respectively, and should be documented in a formal Maintenance Plan that identifies the maintenance tasks to be performed and the resources required in the performance of those tasks. Maintenance resources can include any combination of the following:

1. Labor hours (graded by skill type and level; special training as required)
2. Instruction manuals and support software (e.g., diagnostic programs for computerized automatic test equipment)
3. Tools (common and special)
4. Test equipment (common and special)
5. Facilities (common and special)
6. Replacement materials, parts, and assemblies
7. Consumables and depletables (fuel, lubricants, etc.)

The proper selection of these resources is essential if the product is to be maintained in the way that it was designed to be maintained. The most critical elements for making maintenance effective are the first two resources on the list—properly trained technicians and properly written instruction manuals. The planning tools that we are going to discuss in this chapter will enable technical writers to prepare manual drafts, but the designer should also be personally involved, at least as a reviewer of those drafts, to assure that the procedures take full advantage of maintainability features incorporated in the design and to accommodate special maintenance needs.

Technician training is another matter. Most products are not complex enough to demand specially trained maintenance technicians but, for a growing variety of products ranging from automobiles to computers to nuclear power plants, manufacturers and contractors are providing technician training in order to avoid customer dissatisfaction and, in some cases, liability litigation. Although there is no need for the designer to involve himself in training activities, he should serve as a consultant to the people responsible for developing training instructions and special educational equipment.

You'll note that a distinction is made, for several of the kinds of maintenance resources listed, between "common" and "special." *Common* refers, as you might expect, to generic items, whereas *special* is applied to items that are unique to a given equipment or task type. Thus, most screwdrivers, socket wrench sets, channel-lock pliers, hammers, and the like, are common items found in every mechanic's toolbox. The military services have found it convenient to establish standardized kits of common tools for maintenance technicians assigned to broad equipment categories such as on-the-road vehicles and small fixed-wing aircraft.

A hydraulic lift is a common facility found in any commercial garage or service station. Although it is somewhat expensive and requires considerable floor space, a lift greatly facilitates performing a large variety of maintenance tasks. An automobile wheel alignment assembly, by way of contrast, is a special facility found in relatively few commercial garages because it can be used to perform only one specific task. (Hydraulic lifts and wheel alignment assemblies are classed as facilities, rather than as tools, because they are much too large to fit into a toolbox, and must be permanently installed within a building.)

To an electronics technician, a multimeter is the commonest piece of test equipment because it is the most broadly useful, but a tester for checking out printed circuit boards is special because it must be designed to accommodate the circuitry of specific board configurations. (However, with the advent of computer-programmable circuit board testers, the degree of specialization is reduced.)

Confusion occasionally arises in distinguishing between parts and materials (Item 6 in the list above). A *material* is any item with one or more dimensions left undetermined until the time of installation. Materials can range from paint, which has no predetermined dimension, to insulated wire, which has only one non-predetermined dimension, its length. (If an insulated wire is cut to final length when manufactured, it becomes a part. However, fixed-length wires are usually manufactured as a component of an interconnection assembly with connectors or other terminations attached.)

6.3 STRUCTURE OF THE MAINTENANCE ORGANIZATION

In planning the maintenance of a product, two guidelines should be followed to achieve all necessary support with optimal expenditure of resources. First, maintenance tasks should be performed as close as possible to the place where the product is operated or stored when not in operation, to minimize down-time. Second, corrective maintenance involving the replacement of components should be predicated on replacement of the faulty failure source at the assembly level that affords the briefest down-time, usually the task with the lowest MTTR. With the product restored to service, the faulty removed item can be repaired as convenient or discarded if that is more cost-effective.

However, the maintenance planner rarely enjoys the luxury of planning the optimal maintenance organization from scratch. Maintenance of a new product must usually be planned to fit an existing organization that constitutes a workable compromise between cost, effectiveness, and convenience for both the manufacturers and the end-users of a broad product line. A two-tier or three-tier maintenance organization is typical for many kinds of products. At the lowest tier, servicing and minor repairs are performed by the end-user in accordance with an owner's manual furnished with the product. No special tools, test equipment, or facilities are required, and the necessary replacement parts and assemblies are readily available from local distributors or by ordering from the manufacturer.

At the second tier, preventive or corrective maintenance normally beyond the capabilities of the end-user is performed by an independent local shop. This level of maintenance generally requires trained technicians and may necessitate the use of special tools or test equipment. Some of the replacement parts and assemblies may have to be procured from specialty houses or the manufacturer's regional warehouse, because of lesser demand rates. The most familiar second-tier maintenance facility is the neighborhood garage for repairing automobiles.

The third tier performs corrective maintenance tasks that require special facilities or specially trained technicians. The dividing line between second and third-tier maintenance is not fixed, and varies in accordance with local circumstances. A large auto service center may repair transmissions on the premises, while a small garage sends them out to a shop that specializes in transmission overhauls. For some types of products, the manufacturer may elect to use his own factory for third-tier maintenance because he already possesses the necessary special resources. Independent or manufacturer-operated regional repair centers may be utilized to provide third-tier maintenance for other types of products.

It must be noted that maintenance organizations are not limited to three tiers—at one time, the U.S. Army designated seven distinct maintenance "echelons" for vehicles. However, these echelons did not actually represent seven organizationally separate tiers; the lowest two echelons consisted of daily maintenance and minor repairs, both being performed by the driver. Today, the three-tier organization structure is employed by all of the U.S. military services, even for the most complex and sophisticated electronic systems. The lowest tier is called the "Organizational" tier. For maintenance support of an aircraft, this tier is frequently termed the "Flightline" tier; for a ground-based radar installation, it may be known as the "On-site" tier. The second tier is called the "Intermediate" tier; a mobile shop can serve as an intermediate-tier maintenance facility. The third tier is generally designated as a "Depot," and may be a regional facility operated directly by a military organization or a factory operated by the manufacturer of the faulty item.

6.4 RESPONSIBILITY OF THE DESIGNER

If the designer has done a good job in tailoring the design to satisfy the qualitative maintainability principles set forth in Chapter 2, it will be relatively easy to fit maintenance tasks into a cost-effective maintenance organization structure. In manufacturing firms that do not maintain a logistics organization (and that includes almost all non-military manufacturers, as well as many of the smaller suppliers to the military), the designer must assume the responsibility for whatever maintenance planning is needed to support the product during its warranty period and throughout a reasonable service life thereafter. In some firms, the marketing department may favor a policy of refusing to support superseded models, thereby compelling their users to purchase new models when units break down. (This sort of strategy raises uncomfortable questions about the ethics of the marketplace.) However, the designer's concern should be to plan for lifetime support of the product, regardless of the marketing approach adopted.

The extent of lifetime maintenance support required by a given product depends on a number of factors generally related to the product's complexity. But, at the very least, every product line should be supported to the extent of stockpiling replacement parts and assemblies unique to each given model until that model is no longer in general use. This rule necessitates estimating the consumption rate for each such part or assembly.

For a given newly-designed part or assembly, the starting point for estimating the consumption rate is the item's failure rate, as predicted by a reliability analysis (see Chapter 5). But you must remember that only those failures attributable to chargeable failures are considered in predicting failure rates; i.e., failures caused by a design deficiency or a defect in its component parts or materials, fabrication processing, or workmanship. The failure rate does not take into account the failures caused by operator or maintenance technician error, by mishandling of the product in shipping, storage under excessive periods prior to usage or under improper conditions, inadequate servicing, or misuse of the product. Therefore, the failure rate must be adjusted by a factor which can be determined only from a history of use of the same type of part in a similar product.

Experience dictates that an additional adjustment factor must be applied whose magnitude depends on the relative size, cost, and commonality of the item. Parts that are small tend to be easily lost; screws and washers, for example, frequently roll off a repair bench. If they are inexpensive as well, little effort is expended in preventing them from being lost or in finding them afterwards. Still another replacement rate factor is the diversion of parts to applications other than the ones for which they had been supplied, particularly to personal projects whose end-items do not show up in finished-goods inventories.

For parts and assemblies used in previous applications, actual consumption rate data derived from inventory records can be used for adjusting failure rate predictions to take into account units expended in replacing parts/assemblies destroyed by nonchargeable failure events or lost because of carelessness, diversion, or theft. Such records are of great value because of a tendency by designers to minimize changes in parts and assemblies used throughout several generations of a given model design, as well as among different models within the same general product line.

For replaceable parts and assemblies intended to be produced for a limited period only, it is advisable to add a generous percentage to the estimated total replacement quantity, in order to avoid production line restart costs that would penalize a possible underestimate of actual consumption. The same approach should be taken with regard to special parts and assemblies used in your product that are procured from outside vendors. A good example of a procured item that may be unique to a given model production run is a printed circuit board assembly. If a board assembly has to be re-ordered when it is no longer in production, the manufacturer may charge as much as five times the original price. But, when calculating quantities of spares to be procured, reorder surcharges must be offset by spares storage costs and the cost of the capital tied up in these spares.

If a procured item is an assembly that can, in turn, be broken down into replaceable parts or subassemblies, consideration should be given to procuring replacement units at both the higher and lower assembly hierarchy levels. It is desirable to bias the higher/lower level mix toward maximizing the quantities of lower-level items, in order to minimize the extent of the non-failed hardware components that will ultimately have to be discarded because they are part of higher-level items that cannot be repaired because spare units of certain critical parts are no longer available. Take the case of a carburetor with a faulty needle valve: it is far less costly to replace the needle valve than it is to discard the rest of the carburetor, provided that a spare needle valve is available.

However, the cost-effectiveness of this multi-level logistics strategy is subject to two major constraints. First, replacing the lower-level item may constitute a greater drain on other maintenance resources because it creates additional access problems or requires special tools, test equipment, or facilities. Second, estimates of the replacement quantities needed at each assembly hierarchy level may be subject to a considerable degree of uncertainity. In the latter situation, it is preferable to provide a margin of safety by shifting the level mix bias a bit closer to the higher-level items, because failed higher-level items can be "cannibalized" as necessary to fill gaps as they occur in the inventories of lower-level components. Spares estimating must be classed as a fine art, but a great deal of guidance is available from the current technical literature (see Appendix A).

6.5 IDENTIFYING THE MAINTENANCE SUPPORT RESOURCES NEEDED

Now that we have learned from Chapters 4 and 5 how to identify all the preventive and corrective maintenance tasks to be performed in order to support the product throughout its useful service life, identifying the maintenance resources needed for providing that support is a straightforward, if time-consuming, chore.

The resources needed for performing each maintenance task identified on the RCM or CMA form by Failure Source number (see Figures 4-6 and 5-2, respectively) are listed on the *MRL (Maintenance Resource List;* see Figure 6-1). The column headings on the MRL form are generally self-explanatory; however, a few comments may be helpful. For Labor Grade entries, a hierarchic coding should be established by organization tier and area of technician specialization. For example, for automotive maintenance, the coding scheme listed in Table 6-1 might be employed.

Rather than listing individually each of the common tools needed for performing a given task, it may be more convenient to group common tools into one or more standard tool kits. (In our example, a single standard kit may suffice the service station attendant, but each of the specialists at the dealer's service facility may require

	NAME	PART NUMBER
ITEM UNDER ANALYSIS		
NEXT HIGHER ASSEMBLY		
SUBSYSTEM		

SHEET OF

FAILURE SOURCE	LABOR GRADE	TOOLS	TEST EQUIPMENT	FACILITIES	REPLACEMENT ITEMS	CONSUMABLES & DEPLETABLES	TIER

Table 6.1 Labor Grade Coding Scheme

A. Roadside (operator)

B. Service station (attendent)

C. Independent (garage)
   1. Mechanic, general
   2. Mechanic, engine/driveline
   3. Mechanic, electrical

D. Dealer's serivce
   1. Service manager/diagnostician
   2. Specialist, engine/carburetor
   3. Specialist, automatic transmission
   4. Specialist, manual transmission
   5. Specialist, electrical
   6. Specialist, suspension
   7. Specialist, ventilation/air conditioning

a different kit.) Then, items in the standard tool kits need not be listed on the MRL, since they will already be available to the technician identified by labor grade.

Common test equipment should be also be included in the standard tool kits, and need not then be listed on the MRL. Certain facilities can be considered as common facilities in the sense that a given organization tier is always equipped with the same basic facilities. Thus, a service station is always equipped with a hydraulic lift, compressed air lines, and pneumatically powered wrenches; these items would be used for a task involving removal and replacement of automobile wheels. Instead of listing each facility individually, it would be preferable to enter a code designation such as the alpha character, B, which denotes "service station" in our labor grade example.

Since one of the items usually replaced in the maintenance task associated with a given failure source is the failure source itself, which has already been comprehensively identified on the RCM or CMA, time and effort are saved by entering a code such as "F/S,"

Figure 6.1 Maintenance resource list (MRL) form.

instead of repeating the complete identification for the failure source as replacement item.

In filling out the MRL, you may encounter tasks which a suitable tool, test equipment, or facility is not available. You have two choices: either design and fabricate the nonexistent item, or redesign the product to eliminate the need. This possibility is, perhaps, the best argument for starting your maintenance planning very early in the product development process.

6.6 INTEGRATING THE PRODUCT INTO THE MAINTENANCE ORGANIZATION

Now that we have identified the resources required to support maintenance of the product, we can take the next step in maintenance planning, and identify the organization tier at which each maintenance task will be performed by entering the appropriate code in the last column of the MRL. The general rule here is to assign a given task to the lowest tier at which the task can be accomplished within the constraints of the labor grade and common resources available at that tier.

In the automotive repair example that we have used repeatedly in this book, the choice of maintenance tier has, for most tasks, already been made by the act of selecting the maintenance resources, since most of the resources are already designated as associated with specific tiers. If the task can be performed by a service station attendant using a standard attendant's kit and a hydraulic lift, there is no need to go to a higher tier. Alternatively, a radical new product design may justify modification of the existing maintenance organization structure—the maintenance planning process is the means of alerting the manufacturer to such possibilities.

6.7 PHASED-MAINTENANCE SCHEDULING

A completed MRL for identifies the resources that must provided for supporting maintenance of the product, but does not tell us when these resources should be used. Preventive maintenance must be properly scheduled in order to be effective. (Corrective maintenance cannot, of course, be scheduled.) The basis for preventive-maintenance scheduling is the completed RCM (Reliability Centered Maintenance) form described in Chapter 4, which lists the tasks to be performed and their frequency of performance. Three types of maintenance are scheduled: servicing; replacement of limited-life components; and condition assessment, including verification of hidden functions.

If the frequency data from the completed RCM form (i.e., the entries in Blocks 1.4, 2.3, 3.2, and 4.2) are plotted on a chart of time or other measure of operating life, in all probability the individual maintenance events will be almost randomly dispersed. If this chart were used as-is for a preventive maintenance schedule, product down-time would be excessive and labor allocation most inefficient. *Phased-maintenance scheduling* is the procedure for modifying this initial schedule to minimize down-time and optimize labor expenditures.

The first step is to separate the servicing tasks from the other preventive maintenance tasks. Servicing tasks are usually performed much more frequently, and tend to be relatively simple and closely associated with operating cycles. In many cases they can and should be performed by the operator rather than a technician. In the case of an automobile, the driver can check the tires, battery, radiator, and crankcase weekly. An aircraft pilot will perform all these kinds of inspections and, in addition, do a quick walk-around the plane before each flight to look for structural damage—he will also exercise the flight controls before taking off.

The second step is to develop a modular interval for condition assessment inspections. This may involve some "cut-and-try" iterations before a best-fit modular interval is found. Each frequency entry from Block 4.2 is treated as a maximum allowable interval for the associated inspection—the inspection can be performed more frequently but not less frequently. For an automobile, a suitable interval module might be 3000 miles—the automobile is lubricated at the end of every module interval (lubrication, in this case, is treated as a condition assessment rather than as a servicing, because of the length of the interval). Also inspected at the end of every interval is operation of the braking systems, of exterior lights, and of the windshield wipers (a hidden function). At the end of every second interval, the crankcase oil is changed; at the end of every fourth interval, the brake linings are inspected and an engine tune-up is performed; at the end of every sixth interval, wheel alignment is checked. Of course, any one of the inspections will be performed earlier if some problem is indicated by condition monitoring.

The third step is to find a modular interval for limited-life replacement tasks in the same way as we did for condition assessment, and compatible with the module adopted for the condition assessments.

The fourth step is to integrate the condition assessment and limited-life replacement tasks into a single schedule. For an automobile, the crankcase oil filter could be replaced every 6,000 miles and the air filter every 12,000 miles.

APPENDIX

Most works on the subject include maintenance planning as part of an even broader topic, logistics. For a number of writers, logistics incorporates reliability, maintainability, and availability to the extent that they affect the support of operating systems. A publication by Blanchard (1) exemplifies this approach, and it is a very useful one because it serves to integrate subjects that are all too often treated in isolation.

An early work by Goldman and Slattery (2) supposedly on maintainability is actually directed to what is now designated as logistics, but can be useful in showing the relationship between the two fields. Life cycle cost analysis, which is a facet of maintenance planning deserving a publication of its own got one recently in the form of a military handbook (3) addressed to Navy needs. However, it does furnish a concise introduction to systematic analysis procedures.

REFERENCES

1. B. S. Blanchard, *Logistics Engineering and Management*, Prentice-Hall, Englewood Cliffs, N.J., 2nd Edit., 1981.
2. A. S. Goldman and T. B. Slattery, *Maintainability: A Major Element of System Effectiveness*, Wiley, New York, 1964.
3. Military Handbook MIL-HDBK-259 (Navy), *Life Cycle Costs in Navy Acquistions*, U.S. Dept. of Defense, 1983.

III
RMA Analysis

In Part I, we learned to express the inherent reliability of an item in terms of failure rates, with the failure rate concept explained by visualizing a series-connected circuit as a chain whose failure is synonomous with the failure of any of its links. The failure rate of the product modeled by the series-connected circuit is then obtained by summing the failure rates of its components.

Part II provided the methodology for analyzing the design of the item to calculate (and improve) its availability, and to plan for its maintenance.

We also learned that some products do not fit this simple model and must be represented as systems of series-parallel circuits. Part III lays out the technique for improving the reliability and availability of a given system beyond those of its basic configuration by adding redundant elements to reinforce critical weak-points.

7

Fundamentals of RMA Analysis

7.1 PRODUCT DESIGN VERSUS SYSTEM INTEGRATION

The designer of a product is not directly responsible for the design of every hardware component of the system at every level down to the nuts and bolts. He is not necessarily responsible for the design of every assembly used in the product. Wherever possible, he utilizes standard items currently available on the market, since he and his company could not hope to match the price and quality of an "off-the-shelf" item they are not already producing themselves. Even for items that must be custom manufactured, a "make-or-buy" evaluation frequently discloses that procurement from an outside source is less costly. Thus, the designer of a relatively simple product like a lawnmower may be considered to be a modest practitioner of *system integration*. In other words, a substantial part of his task is to select off-the-shelf parts, materials, and assemblies and work them into the optimal lawnmower, within the cost and other constraints applicable to manufacturing and marketing the lawnmower. Only the design of the overall product is truly his.

The designers of a communications network are full-time system integrators—other people have to do the worrying about component design, even when the components are equipment items of substantial size. The network designers try to make do with equipment already on the market, although they may find it necessary to oversee the development, by subcontractors, of new equipment with special capabilities or with capabilities beyond those already available.

7.2 MODELLING SERIES-LINKED CIRCUIT COMPONENTS

In Section 1.4 we learned the following approach to predicting the reliability of a product:

1. If the product is made up of a number of functionally interconnected components, and the failure of any of these components results in failure of the product, the product constitutes a system that can be depicted as a chain, with each one of the components comprising one link of that chain. Thus, the chain can be used to represent a circuit, such as the coolant flow circuit of an automobile engine, or one of its mechanical assembly circuits, such as the driveline from the engine to the powered wheels.
2. Each of the components of the system can be assigned a failure rate that expresses in quantitative terms an inherent physical characteristic determining the vulnerability of the component to imposed functional and environmental stresses.
3. The sum of the failure rates for all of the components equals the predicted failure rate for the system; the reciprocal of this total failure rate equals the MTBF (mean time between failures) for the system.

The chain model is shown in block-diagram form in Figure 7-1. Each block represents one component; each block is marked with the reliability value (failure rate) of the component it represents. The lines between the blocks designate the interconnections between the components. It is essential that the interconnections shown between the blocks accurately represent actual functional interconnections between components so that the block diagram can serve as a reliability model of the system for MTBF and availability predication purposes.

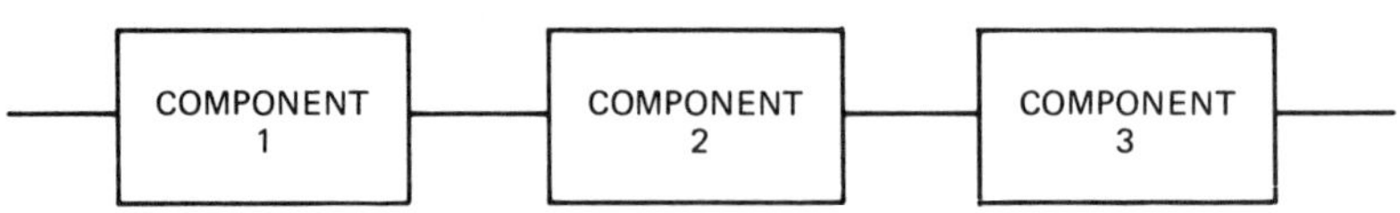

Figure 7.1 Reliability Model – Series-Linked Circuit.

The block-diagram reliability model is analogous to the circuit diagrams used for evaluating the electrical circuits. In the terminology common to electrical analysis, the components represented by blocks in the diagram shown in Figure 7-1 can be described as being *series-linked*.

7.3 MODELLING PARALLEL-LINKED CIRCUIT COMPONENTS

Let's complicate the system shown in Figure 7-1 by connecting an added component in parallel to one of the existing components—this gives us the model shown in Figure 7-2. What real-world situation does this model represent? Well, it's analogous to the electrical circuit we talked about in Section 1.8, in which two resistors were connected in parallel. If one of those resistors burned out, the circuit would still operate, albeit with somewhat degraded performance, so long as the other resistor continued to function. A reliability model with two or more components in parallel indicates that the circuit thus depicted will continue to perform its function so long as any one of the components is operable. And just as the equivalent resistance of a parallel combination of resistors is not equal to the individual resistance of any of its members, so the equivalent failure rate of a combination of components in parallel is not equal to the individual failure rate of any

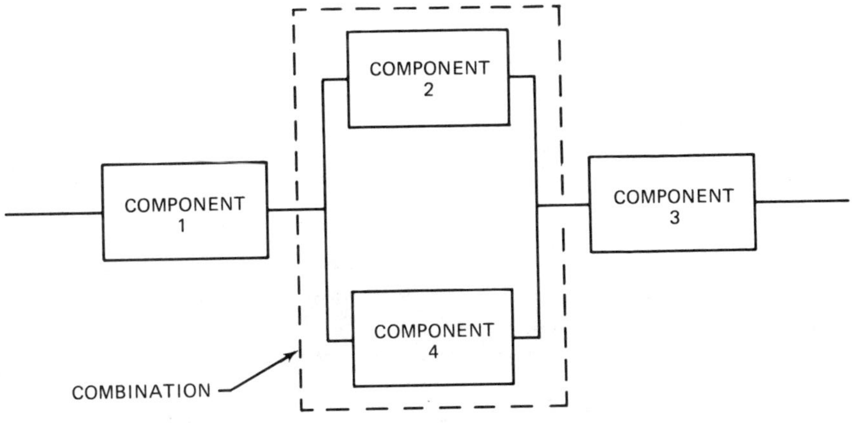

Figure 7-2 Reliability Model – Series/Parallel Circuit

of its members—in both cases, the equivalent value for the combination is less than the individual value of any member.

Equations (1) thru (5) below for equivalent failure rates or MTBF's are derived from the fundamental exponential reliability prediction equation given in Section 1.4. We won't go into the several references where the details can be found—we'll just have to assume they are correct.

The equation for the equivalent MTBF of two components in parallel is as follows:

$$\frac{1}{\lambda_e} = \frac{1}{\lambda_1} + \frac{1}{\lambda_2} - \frac{1}{\lambda_1 + \lambda_2} \tag{1}$$

Where λ_e is the equivalent failure rate for the combination, $1/\lambda_e$ is its equivalent MTBF, and λ_1 and λ_2 are the failure rates of the members of the combination. (The equation is stated in terms of the reciprocal of the equivalent failure rate in order to arrive at the simplest expression.)

If the failure rates of the two components are equal, the equation for the pair of components connected in parallel reduces to:

$$\text{MTBF} = \frac{3}{2\lambda} = \frac{3}{2}\theta \tag{2}$$

Where θ is the MTBF of each of the components.

The equation for the equivalent MTBF of three components in parallel is:

$$\frac{1}{\lambda_e} = \frac{1}{\lambda_1} + \frac{1}{\lambda_2} + \frac{1}{\lambda_3} - \frac{1}{\lambda_1 + \lambda_2} - \frac{1}{\lambda_1 + \lambda_3} - \frac{1}{\lambda_2 + \lambda_3} + \frac{1}{\lambda_1 + \lambda_2 + \lambda_3} \tag{3}$$

If the three components have equal failure rates, the equation reduces to:

$$\text{MTBF}_e = \frac{11}{6}\theta \tag{4}$$

As you can see, the equivalent MTBF equation gets messy very quickly as the number of parallel components is increased, unless all components in the combination have the same failure rate. But when they all have the same rate, the general equation for the equivalent MTBF of a combination of "n" components in parallel reduces to:

$$\text{MTBF}_e = \sum_{i=i}^{n} \frac{1}{i \times \lambda} = \sum_{i=i}^{n} \frac{\theta}{i} \qquad (5)$$

7.4 IMPROVING CIRCUIT RELIABILITY THROUGH REDUNDANCY

Since the effective failure rate for a combination of components connected in parallel is less than the failure rate of any one of its members, it appears that we can improve the reliability of a system containing a "weak" component (i.e., a component with a relatively high failure rate) by adding a like component paralleling the weak one. However, improving reliability through *redundancy* is a method subject to several restrictions. As we have seen, in the example of the added resistor, the effective performance parameter value for the combination is not the same as for the original component alone. If that component is a resistor rated at 40 ohms, adding an identical resistor results in an effective resistance of 20 ohms for the combination. A failure of either one of the resistors in the combination will bring the effective resistance back to 40 ohms. Unless this circuit is capable of operating within acceptable tolerance limits with the resistance at either 20 or 40 ohms, the redundancy method cannot be used.

The designer must carefully evaluate the possible consequences of every failure mode of the components involved in the redundancy application. Suppose that the redundant resostor could fail in a closed-circuit mode, as well as in the more common open-circuit failure mode. Then the addition of the redundant resistor would not eliminate the possibility of a short-circuit condition.

Another thing to bear in mind when thinking about using the redundancy method is the fact that the reliability improvement thereby gained presupposes prompt replacement of every redundant component that fails, since system reliability is reduced until the replacement has been made.

Up to this point we have been looking at the redundancy method primarily as applied at the lowest assembly hierarchy level, which is the level of the simple assembly or circuit comprised of elemental components. But the redundancy method can be applied at any level. We'll learn how to select the optimum level(s) when we get to Chapter 8.

7.5 APPLYING THE REDUNDANCY TECHNIQUE

To see how redundancy is applied in a real-world situation, let's look at a ground-based radar installation with electronic equipment grouped into two identical channel segments, component for component; these two parallel segments feed into a single electromechanically rotated radar antenna. The two electronic channel segments function totally independent of one another; each of the two segments plus the antenna used in common comprises a transmission channel. This configuration is typical of air traffic control radars operated by the Federal Aviation Administration (FAA), and can be represented by the block diagram in Figure 7-3. A single rotating antenna is employed because its reliability has been found to be comparable to the effective reliability of the two parallel electronic channels combined. (A practical rule-of-thumb derived from the relevant mathematical relationships holds that circuit reliability is optimal when all series-connected components are equal; enhancement of series-linked circuit reliability by improving the reliability of any one component above the common level tends to conform to the law of diminishing returns.)

The radar system is operated around-the-clock, with Channel Segments A and B on-line alternating in 12-hour shifts. The off-line period for each segment is utilized for condition-assessment and other preventive maintenance chores. In the event that a segment fails while it is on-line, the other segment is immediately

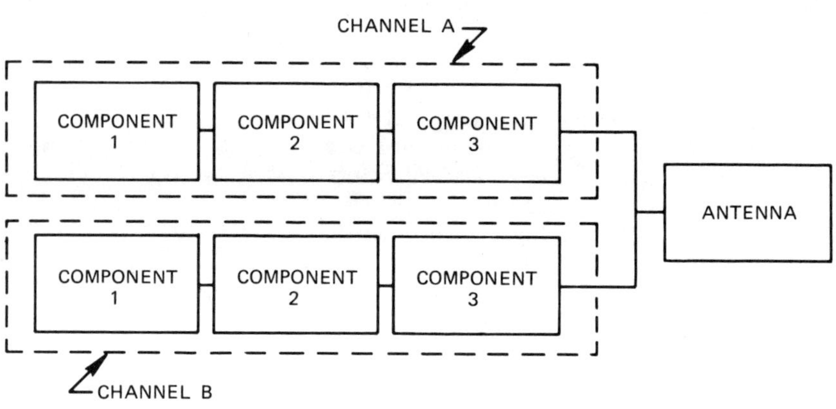

Figure 7.3 Reliability Model of Ground-Based Radar Site

switched on-line, and corrective maintenance is performed on the failed segment, now in the standby-mode.

From the reliability analysis standpoint, it is of primary importance to note that a channel segment failure is not necessarily a radar system failure. The cardinal rule in classifying failures is to evaluate the failure event against the mission definition. The mission of this radar system is to generate a continuous radar beam of specified characteristics—since the segment failure did not cause the beam to be interrupted or degraded beyond allowable tolerance limits, a failure at the system level has not occurred—system availability has not been reduced. If, however, restoration of the failed channel segment requires a significant period of time to accomplish, system reliability will be decreased to the non-redundant configuration level during that period.

Remember that a reliability value is a prediction—it represents an expectation, calculated in accordance with the laws of probability, and can be applied only for the period of time during which the external stresses that affect the reliability of the system can be expected to remain stable. Actually, the prediction is 100% valid only for the moment that begins the given period, because the item will undergo internal changes in response to those stresses and to various degradative agents (e.g., wear and corrosion) acting over a period of time that are continually operating to reduce its reliability. The purpose of preventive and corrective maintenance is to counter this loss of reliability.

Returning to the radar example, you will recall the statement that a channel segment failure would not cause a system failure because of the capability to switch from the failed segment to the standby segment. But this switchover requires a finite time period to accomplish. The reliability specification for operation should state the maximum allowable duration of interruptions for switchover that can be tolerated without constituting a system failure. If human response to a channel segment failure is too slow or too uncertain to satisfy this requirement, the design will have to incorporate provisions for automatic detection and switchover.

The switch-over duration limit can also affect the design of systems with a lengthy start-up time. Suppose that our radar system usually draws electrical power from a large municipal plant but includes a standby diesel-driven electric generator for protection against interruption of municipally supplied power. This diesel-electric unit requires five minutes to come on-line, but the maximum allowable duration for interruptions of radar operation is only two minutes. In order to meet the specification, it will be necessary to keep the diesel in "hot-standby" mode; i.e., to keep the diesel in continuous operation, unless the specification is relaxed or other provisions are made to obtain power as needed.

APPENDIX

Most of the references listed in the appendix to Chapter 1 discuss reliability modeling and prediction for complex systems. Analysts employing data taken from Military Handbook MIL-HDBK-217 (Reference 3 in the appendix to Chapter 1) will be interested in the modeling and prediction techniques described in MIL-STD-756 (Reference 4). However, MIL-STD-756 approaches the analysis of circuit reliability block diagrams by assigning values expressed in probability of success/failure units to the individual blocks, instead of failure rates as used herein. At the circuit level, reliability is again expressed in terms of probability of success/failure (instead of MTBF), a unit which is not directly compatible with units for measuring maintainability and availability.

The Navy publication referenced below takes a more practical approach to the application of redundancy in circuit design. Here again, reliability is expressed in terms of probability of success/failure, but relationships for the more common circuit block diagrams are translated into MTBF equations. These include derivations of the equations given in Section 7.3. In addition, and of particular interest, are techniques for dealing with reliability analysis of several commonly utilized methods of bypassing the redundant-component circuit design problems discussed in Section 7.4. (Among these methods is the alternate-segment switching approach described in Section 7.5. Be warned that the analysis technique employs probability of success/failure equations, but the solutions of such equations can be converted to failure rate or MTBF units.

REFERENCE

Handbook, NAVWEPS 00-65-502, *Handbook of Reliability Engineering*, U.S. Navy, 1964.

8
Analysis of Complex Systems

8.1 MODELLING COMPLEX SYSTEMS

The analysis technique described in Chapter 7 is adequate for handling most products, since most products can be represented as groups of series-linked components, with only an occasional parallel connection. But there are other kinds of products which present a higher degree of series-parallel complexity, most notably communications *networks* that interconnect equipment located at widely dispersed sites. These networks are required to exhibit extremely high system-level reliability and availability because of the nature of the missions they perform.

In designing this kind of product, tough system-level reliability and availability specifications are usually met by incorporating equipment redundancies in the most cost-effective manner, subject to the restriction that the equipment must be purchased as off-the-shelf hardware building blocks—you can only buy what the manufacturer already produces and sells. (Typically, hardware manufacturers will either refuse to customize their circuit designs by replacing commercial-grade components with high-reliability equivalents, or will charge exorbitant prices for having to disrupt routine procurement and assembly procedures.)

The redundancy approach to system reliability improvement offsets the cost of the additional components, and of the additional space, weight, and power needed for them, with certain advantages. Remember the dual-channel radar configuration mentioned in Section 7.5? Each channel segment is redundant to the other. When a given segment is in the on-line mode, condition assessment of its critical components is prohibited because it would interfere with operation. But, with the dual-channel configuration, these inspections can be performed alternately on each segment during its scheduled standby mode.

8.2 INTRODUCING THE COMMUNICATIONS NETWORK EXAMPLE

Figure 8.1 depicts a hypothetical microwave telecommunications network linking ground-based radar sites that comprise an air traffic control system. The RMA analysis of this network is quite complex and lengthy, but it will provide us with an opportunity to exercise just about every concept discussed in previous chapters. The Communications Network which serves as our example interconnects three kinds of sites:

1. A single Control Center, which coordinates the operation of the Communications Network. The Control Center originates messages to and receives messages from all of the radar sites.
2. Radar sites, each of which tracks aircraft passing through the airspace sector monitored by that site, and provides an aircraft with position and weather data upon request. Each radar site originates messages to and receives messages from the Control Center and other radar sites in the Network. Most radar sites also relay messages between other other radar sites and the Control Center, and between the other radar sites themselves.
3. Repeater sites, which only serve to relay messages between radar sites, and between radar sites and the Control Center.

Network communications links between sites are comprised of 72 microwave radio channels, 36 operating in each of the two directions of transmission. As Figure 8.1 shows, most radar sites are "intermediate radar sites," i.e., radar sites linked (via repeater sites) with two adjacent radar sites, one on either side. However, some of the radar sites in Figure 8.1, specifically Sites A, G, J, T, X, and Y are linked with only one other radar site because they are at the end of a series of sites—these are "terminal radar sites." By way of contrast, Sites B, M, R, and W are each linked with more than two radar sites—these are "junction radar sites." Similarly, the Control Center is linked with seven radar sites.

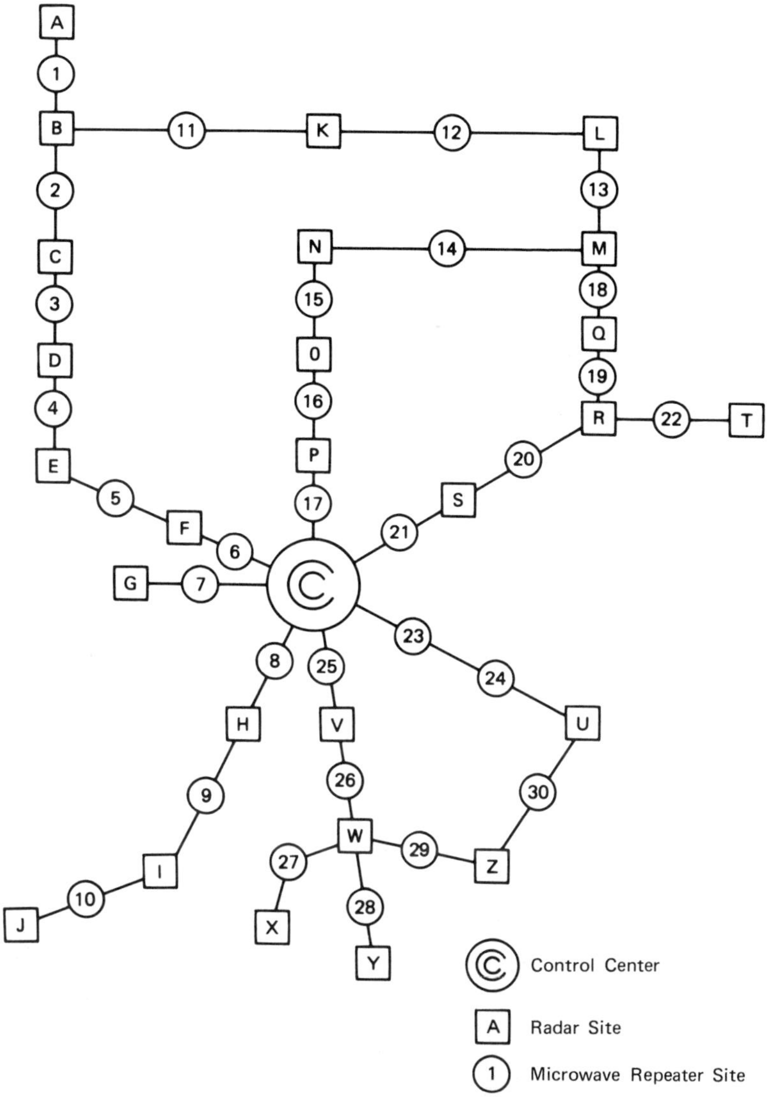

Figure 8.1 Communications Network configuration.

8.3 SPECIFYING RMA FOR THE NETWORK EXAMPLE

The customer has specified the RMA (reliability/maintainability/availability) requirements set forth below for the Communications Network.

8.3.1 Link-Level Requirements

"Link-level requirements shall address the complement of equipment installed at any two adjacent, microwave-interconnected radar sites (i.e., radar sites interconnected directly or via repeaters), plus repeater equipment, for the purpose of transmitting messages being relayed by the link or originating or terminating within either of the two included radar sites.

For a link between two intermediate radar sites, the link equipment complement shall be equivalent to the equipment comprising one complete intermediate radar site plus that of all the repeater sites included in the link.

A link failure shall be defined as an interruption of message transmission in either direction for more than one minute. Any instance wherein fewer than 30 channels are available in each of the two directions shall constitute an interruption of message transmission, except that only six channels shall be required within a link including a terminal radar site."

8.3.1.1 Link MTBF

"The effective MTBF of any link within the Communications Network shall be not less than 600 hours, where effective MTBF is a function of interruptions of communication across the link rather than failures of individual components of the link."

8.3.1.2 Link Availability

"The availability of any link within the Communications Network, as measured on an annual basis, shall be not less than 99.50%. A link shall be unavailable from the initiation of message transmission interruption until transmission has been restored."

8.3.1.3 Link MTTR

"No link within the Communications Network shall have an MTTR exceeding 4.0 hours."

8.3.1.4 Link M/max/ct

"No link within the Communications Network shall have an M/max/ct at the 95th-percentile exceeding 6.0 hours."

8.3.2 System-Level Requirements

"System-level requirements shall address the complement of equipment required for communication between the Control Center and any single radar site. A system failure shall be defined as an interruption of message transmission in either direction for more than one minute. Any instance wherein fewer than 30 channels are available in each of the two directions between the Control Center and any single radar site shall constitute an interruption of message transmission, except that only six channels shall be required within that segment comprised of a communication link including a terminal radar site. In those cases where more than one path (sequence of links) interconnects the Control Center with a given radar site, a system-level failure has not occurred so long as the minimum number of channels are available along any one of the paths."

8.3.2.1 System MTBF

"The effective MTBF for communication between the Control Center and any radar site shall be not less than 200 hours."

8.3.2.2 System Availability

"System availability, as measured on an annual basis, shall be not less than 97.50%. A system shall be unavailable throughout the duration of a system failure until message transmission has been restored."

8.4 VERIFYING COMPLIANCE WITH RMA REQUIREMENTS

Specification requirements are ineffective unless the procurement contract includes provisions for evaluating compliance (and for enforcing corrective action when the design is found non-compliant). The wording of the specification requirements above permit evaluation of compliance with all but two of the RMA requirements by means of a prediction analysis; the two exceptions (Paragraphs

8.3.1.2 and 8.3.2.2) require verification by measurement of the availability of the installed network over a one-year span. Nevertheless, the likelihood of compliance with those latter two requirements can be evaluated by prediction analysis, assuming that the designers have succeeded in avoiding any communication failures caused by adverse atmospheric conditions.

The reason for the two exceptions is that they encompass the reliability and availability not only of hardware portions of the system but, also, of the intervening microwave propagation paths. Since performance over a microwave path is, in part, a function of atmospheric and other conditions outside the control of the designer, he is faced with the task of predicting these atmospheric variations and designing the system to cope with the severest conditions anticipated. (This is a situation where a functional performance requirement is hidden within an availability requirement.)

8.5 THE LOGIC BEHIND THE RMA SPECIFICATIONS

In reviewing the RMA specifications for the Communications Network, you may wonder why the customer found it necessary to impose design requirements that seem to overlap. Reliability, maintainability, and availability limits are all applied at the link level; since availability is a function not only of reliability and maintainability, we appear to be saddled with an excess of requirements for the link. At the system level, we are again subject to reliability and availability requirements—why is it necessary to specify these at both levels? Let's try to figure out what the customer had in mind.

8.5.1 MTBF and Availability at Link and System Levels

The MTBF and availability requirements address both link and system levels because continuity of communication is essential at both levels, but for different reasons. The radar sites each monitor a three-dimensional sector of space through which air traffic routes pass. As an aircraft traverses a succession of adjacent sectors, responsibility for tracking the aircraft is handed over to each successive radar site along its route. Communication at the link level is particularly critical when an aircraft being tracked passes from one radar site's sector into that of an adjacent site.

Communication at the system level serves two kinds of centralized functions: (a) directing responses to abnormal operating situations, especially to peaks in air traffic volume that overload the air traffic controllers; and (b) compiling performance and maintenance data for identifying problem areas that require attention, and for supporting statistical evaluations leading to the development of improvements in operating procedures and hardware/software resources. (The Control Center also makes use of the Communications Network to implement Network message traffic priority assignments.)

MTBF and availability requirements are specified separately for the link and system levels because they are intended to satisfy different needs and to address different hardware configurations—the site-to-adjacent-site link at the lower level and the Control-Center-to-remote-site interconnection at the higher level. Inherent MTBF and availability at the higher level are dependent only in part on the inherent characteristics of the lower-level configuration; the rest of the dependency stems from the effect of adverse atmospheric conditions. Note also that reliability and availability requirement values are necessarily lower at the system level than at the link level because the aggregate failure rate of a much greater number of component elements must be considered.

Since failures of certain types of parts and equipment will not affect the inherent reliability of a system, particularly those items incorporated for convenience or condition-monitoring, the components of the system must be screened to eliminate non-critical items from the system analysis; i.e., those items whose failure would not impair the ability of the Network to perform any of its missions. It should also be borne in mind that many of the items installed at the the various radar sites are not a part of the Communications Network and should not be included in the analysis; these items are utilized for performing system missions such as aircraft detection and tracking (radar), communications between ground-based sites and aircraft, weather monitoring, flight plan data storage and retrieval, etc. The Communications Network is only one of several major subsystems comprising the overall air traffic control system.

One more thing—the reason for the one-minute grace period for message interruptions at either the system or link level is to screen out tolerably brief transient communications interruptions caused by unusually severe atmospheric conditions. However, one minute is too short a time to rely on human operator detection of equipment failure and switch over to redundant units—automatic detection and switchover is thus made mandatory.

8.5.2 MTBF—Link Level versus System Level

Because some of the radar sites are interconnected with the Control Center by more than one path (i.e., via more than one series-linked group of sites), the degree of criticality of a failure occurring in a given link depends on the extent of the resulting interruption of message traffic. The system MTBF requirement (Paragraph 8.3.2.1) penalizes those failures that interrupt communication with sites connected to the Center by only one path, but not those failures that do not disrupt all of the paths in a multi-path interconnection of a site to the Center. Consequently, failures at junction sites that interrupt two or more paths may be particularly damaging to system reliability and availability.

8.5.3 MTTR and M/max/ct versus Availability

Since the inherent availability of a product design can be varied by changing either its inherent reliability or maintainability, the designer can overcome an inadequate availability situation with whatever combination of reliability and maintainability improvements he finds most feasible and cost-effective. Now, if the customer had been an advocate of the *black-box specification* philosophy, he would have specified only the availability parameter. This philosophy is predicated on the assumption that the supplier is the more knowledgeable party in designing the product to fit the customer's needs, and should be given maximum leeway to exercise his expertise. In this approach, the specification is limited to defining the application—then, if the design fails to meet the specification, the supplier can be held fully responsible. Conversely, if the specification goes beyond a definition of the application and, thereby, limits the designer's options, the customer has expropriated design prerogatives at the risk of assuming responsibility for any errors or shortcomings introduced by those additional requirements.

The black-box approach is founded on a concept analogous to that of the RCM methodology discussed in Section 4.1, which requires that no preventive maintenance task be performed unless it is supported by a valid justification, thereby reducing preventive maintenance activities to the essentials. In the same way, the black-box approach permits no design provisions that cannot be justified as essential to the application.

In following the black-box philosphy, however, our customer has made a careful distinction between availability requirements and those kinds of design requirements that can readily be validated as soon as a prototype unit has been fabricated. Confirmation of availability (and reliability) prediction analyses must await lengthy testing or in-service operation of the system. By the time that conclusive data has been compiled, it will probably be too late to correct whatever design deficiencies may have been found.

Fourtunately, inherent maintainability characteristics can be tested as readily as, say, performance capabilities or structural strength. Consequently, the customer finds it advisable to specify MTTR and M/max/ct for critical items, which gives him positive control over one of the two factors determining availability. Also, since MTTR and M/max/ct encompass non-critical items as well as critical items, the customer has effectively limited the technician manhours that must be expended in maintaining the equipment. Although these maintainability specifications do not eliminate the possibility of some equipment items requiring extremely lengthy repair times, they do set very stringent constraints on the aggregate of the repair times for the system and, thus, on the manpower resources that will be required for supporting the system throughout its service life.

One other statement must be made about specifying reliability and maintainability requirements: The MTBF of each equipment item depends on its particular circuitry design, but the MTTR is largely governed by general design practice and should be uniform for all components of a system.

Our customer's RMA specifications illustrate twin objectives: (1) assuring that the product will be available whenever the user wants to use it and for as long as he wants to use it; and (2) minimizing the cost of maintaining the product over its service life period. (Note that the MTTR and M/max/ct limits are addressed to the single-channel level, although an outage of a single channel will not necessarily result in the failure of a link. Therefore, it is obvious that the maintainability specifications are directed more toward reducing maintenance cost than down-time.)

8.6 THE TOPOLOGICAL DIAGRAM MODEL

In modelling the Communications Network, systems integration designers will begin with a simple *topological diagram* model; i.e., a diagram that is limited to identifying the major elements of the network and how they are interconnected. Figure 8.1 is just such a model—its layout may or may not indicate the actual geographic layout of the network, and even if it does follow the geographic layout, it is probably not drawn to scale. But the topological diagram model is the necessary first step in analyzing the system.

8.7 THE SITE CONFIGURATION

The configurations of the intermediate radar and repeater sites are shown in Figure 8.2. (Each box shown in Figure 8.2 represents a single entity (monolithic package) of an equipment item.) The repeater site is symetrical in that the 36-channel configuration for one direction is the mirror image of the configuration for the other direction—transmitted messages traverse a repeater via a Receiver, Repeater Amplifier, and Transmitter, in that order. Radar sites are partially mirror-image symetrical: All incoming messages pass, in turn, from a Receiver through MUX Components A and B on their way to a Switching Panel—all outgoing messages travel from a Switching Panel through MUX components A and B to the Transmitter. Relayed messages, which do not originate or terminate in one of the radar site's message terminals, pass from one of the two incoming Switching Panels through the input and output stages of the Relay Amplifier to the corresponding outgoing Switching Panel.

Each radar site is equipped with six message terminals, consisting of three telephones, two teletype machines, and a meterological station that transmits locally measured ambient temperature, barometric pressure, relative humidity, wind velocity and direction. Except for the telephone, each of these terminals is equipped with a buffer device for temporary storage and retransmission of messages during peak traffic periods when communication channels are available only for higher priority messages. As indicated in Figure 8.2, the message terminals and their associated buffers are excluded from the Communications Network model because the Network is defined as including only those equipment items whose function is to carry messages, not generate or receive them.

The Control Center is divided into seven segments, one segment for each of the seven groups of sites with which it is directly linked (see Figure 8.1). Each segment is comprised of an equipment complement identical to that of an intermediate radar site. However, site-to-site messages relayed via the Control Center actually enter the Control Center at one segment's Receiver and exit at another segment's Transmitter. In terms of equipment complement for message relay functions, a combination of any two segments of the Control Center is equivalent to an intermediate radar site; the only difference between them is that each segment includes its own set of message terminals, which are not a part of the Communications Network. Similarly, that portion of a junction radar site which interconnects two intermediate radar sites is itself equivalent to an intermediate radar site.

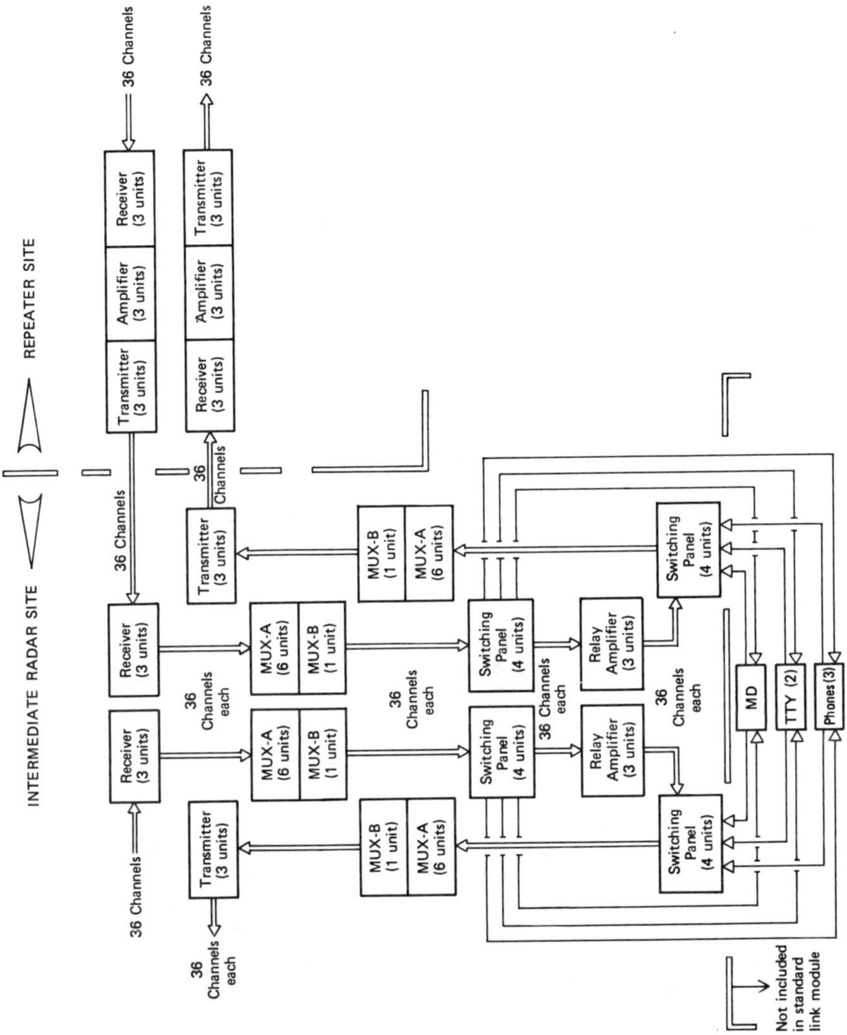

Figure 8.2 Entities comprising standard link module.

A terminal radar site is another matter, however. Since it does not relay messages, a terminal site is linked to the rest of the Communications Network only for messages originating or terminating on-site. This requires the minimal equipment complement—one Receiver entity, one Transmitter entity, and two each MUX Components A and B and Switching Panel entities.

8.8 EQUIPMENT ITEM CONFIGURATION—THE UNIT MODULE

Each one of the Switching Panel entities at a radar site or Control Center segment is made up of four 9-channel units interconnected to accommodate up to 36 channels. The 9-channel units are identical modules, which not only enables the designer to tailor Switching Panels to communications networks of different capacities, but facilitates manufacturing, shipping, and installation. Most important to us, however, is the fact that a failure in a Switching Panel entity can be isolated to the faulty unit and corrected by replacing that unit, instead of replacing the entire entity.

Similarly, the 36-channel Transmitter, Receiver, and Repeater Amplifier entities are each made up of three modular, separately replaceable 12-channel units, and the MUX Component A entity of six 6-channel modular, separately replaceable units. Only MUX Component B is a monolithic 36-channel entity; here the lowest-level replaceable unit is the entire entity.

The Relay Amplifier entities are each comprised of three modular, separately replaceable 12-channel units. The differ from the other equipment item entities in one important respect, however. Whereas the incoming and outgoing channels are served separately by the other equipment item entities, the Relay Amplifier entity accepts and outputs the same 36-channel transmission stream.

Unlike the Switching Panel and Relay Amplifier entities, the 36-channel Transmitters and Receivers do not actually handle 36 physically separate signal channels—they transmit and receive a single broadband frequency signal which embodies the 36 channels as segments of the overall broadband frequency spectrum.

The MUX is a frequency-division multiplexer/demultiplexer which converts output signals from the Switching Panel channels to narrow frequency bands that, in combination, make up the broadband-frequency output signal from the transmitter. Conversely, the MUX splits the incoming broadband signal from the Receiver into narrow frequency band segments corresponding to each channel, and restores these channel signals to the form suitable for the Relay Amplifier or the on-site message terminals, as applicable.

8.9 LIMITING THE NUMBER OF CHANNELS

The message carrying capacity of a link is limited to 36 channels in each direction, which would not be anywhere near adequate for handling the maximum possible message traffic that could be generated within the Network. However, an analysis has predicted that message traffic will occupy less than 30 channels of any link in either direction during 99.9% of system operating time; the probability of an overload occurrence will be significant only when channels become unavailable because of equipment failures. This explains why the customer has specified such a strict availability requirement.

(You needn't wast time trying to understand in detail the telecommunications technology suggested by the description of our example—it's not intended to describe an actual system but to illustrate the kinds of complexities and judgmental decision-making that may be encountered in analyzing a real system.)

8.10 ASSIGNING THE MESSAGE TRAFFIC PRIORITIES

Messages are directed by Switching Panels into available transmission channels. All channels are interchangeable—no channels are dedicated to specific message sources or terminals. An automatic testing procedure coordinated by equipment located at the Control Center continually checks each channel within each link, and alerts the Switching Panels affected whenever any channel has become unavailable because it has been put into use or interrupted by an equipment failure.) Whenever no channels are available, the priority assignment procedure programmed into each outgoing Switching Panel is actuated automatically—buffers in Relay Amplifiers and in all of the on-site message sources except the telephone are activated, in order of priority, to record the outgoing messages and defer transmittal until channels become available. Unlike the other on-site message sources, the telephone has no buffer. However, telephone communications are given a high priority and, in the case of an emergency, the telephone user can manually override the automatic priority system.

8.11 SIMPLIFYING THE TOPOLOGICAL NETWORK DIAGRAM

By now, you probably have some idea as to how complicated a system analysis can get. The right approach, as always, is to simplfy.

The first step toward simplification has already been taken by designing the various sites to be configured from the minimum feasible number of modular equipment item building blocks. But, at the system level, we are faced with the complication of accounting for redundancies constituted by the alternate paths incorporated in the Network (see Figure 8.1). The alternate paths are clearly seen in Figure 8.3, which shows the system of Figure 8.1 redrawn (in strict topological schematic form) as a network of "legs," rather than of individual sites.

Each leg is comprised of a series of links terminated on either end by the Control Center, by a junction site, or by a terminal site. Most of the paths linking one site to another or to the Control Center traverse a series of legs. To distinguish between a leg and a path, look again at Figure 8.3—the route from Site B to the Control Center can follow any one of three paths—via leg II along, via Legs III and IV, or via Legs III, V, and VI. With three-path redundancy, the reliability and availability of the communication link between Site B and the Control Center may prove to be greater than that of the link between Site J and the Control Center, even though the single available path of the latter traverses only two repeaters and two radar sites.

8.12 THE ANALYSIS SEQUENCE

Now that we have developed a simplified approach to modelling the Communications Network, we can get on with the job of actually performing the RMA analysis to determine whether or not our proposed system meets the requirements specified by the customer. If the system does not meet all six requirements, we can go further—we can add redundant elements to the basic configuration, in optimal sequence, until it does.

Failure rate and repair time ratings for individual equipment items have been derived in the manner described in Chapter 1 and provided to us, the system designers, by the manufacturers of those items (see Table 8.1). Presumably, they have derived the ratings from extensive test or field data. The repair time ratings could be obtained either as single-run-through measurements made in the Lab or as MTTR values calculated from a number of actual repair actions. (It should be understood that an MTTR for a system component constitutes the nominal repair time for that component, consequently, the system MTTR can be calculated as the mean of the MTTR's of its components.)

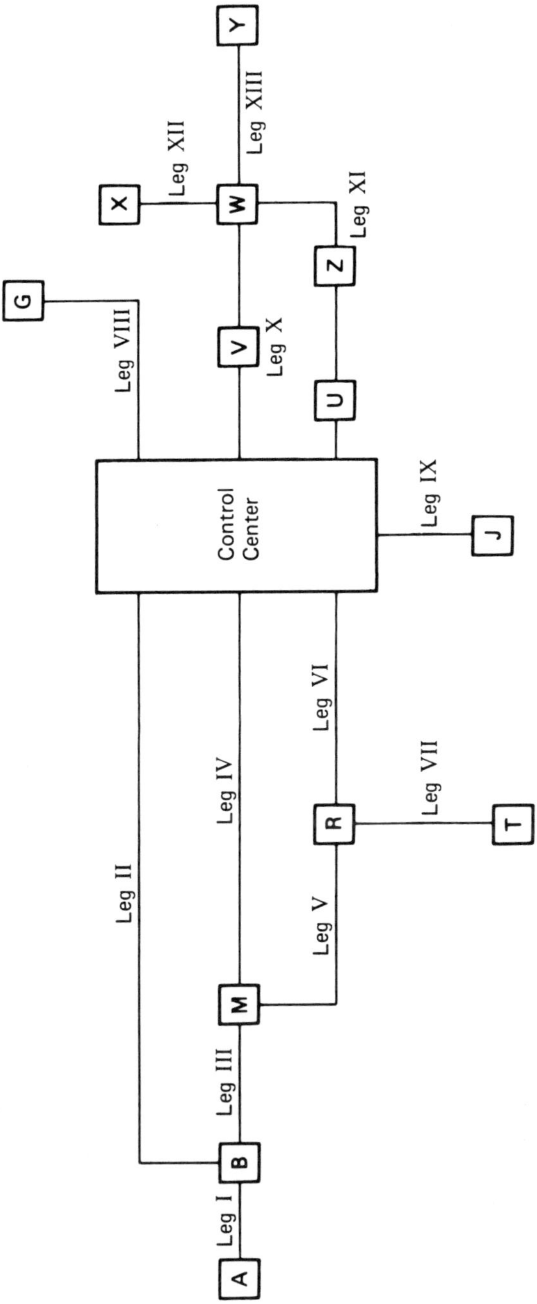

Figure 8.3 Communications Network block diagram by legs.

Table 8.1 Reliability and Maintainability Ratings Assigned to Link/System Building-Block Equipment Items

Equipment items	Quantity of channels per unit	Channel failure rate (FPMH)	Unit failure rate (FPMH)	Entity repair time[a] (hours)	Qty. of units per entity[b]	Entity failure rate[c] (FPMH)
Receiver	12	1.1	13.2	3.8	3	39.6
Transmitter	12	1.2	14.4	3.8	3	43.2
MUX Component A	6 6	98.0 98.0	588.0 588.0	3.6 1.8	6 (5 of 6)	3528.0 36.9
MUX Component B	36	0.9	32.4	2.1	1	32.4
Switching panel	9	13.3	119.7	3.8	4	478.8
Amplifier, relay	12	6.2	74.4	4.8	3	223.2
Amplifier, repeater	12	1.0	12.0	5.6	3	36.0

[a] Active item repair time with repair accomplished by replacing individual faulty unit. For MUX Component A, second value is Einhorn equivalent downtime for item—used only for calculating link availability.
[b] Minimum quantity of units required to be operating at any given moment in order to meet 30-channel link message capacity requirement for each direction.
[c] Aggregate failure rate for minimum quantity of units. For MUX Component A, first value is the sum of the failure rates for 6 units, and second value is Einhorn equivalent for 5 of the 6 units.

From the data in Table 8.1 we will perform the analysis in the following order:

1. Calculate the link MTBF, availability, MTTR, and M/max/ct values predicted for the design, and compare them with the corresponding requirements specified under Paragraph 8.3.1.
2. If any of the link-level requirements is not satisfied by the predicted values, determine which redundant elements are to be added to the system design, in optimal sequence, to eliminate the deficiencies found.
3. Calculate the MTBF and MTTR values for each leg.
4. From the leg parameters, calculate the system MTBF and availability values, and compare them with the corresponding requirements specified under Paragraph 8.3.2.
5. If either of these system-level requirements is not satisfied by the predicted values, determine which redundant elements are to be added to the system design, in optimal sequence, to eliminate the deficiencies found.

The reason for remedying any deficiencies found at the link level (Step 2 above), before moving on to the leg and system levels is that improvements at the link level will also improve the higher level characteristics, but not vice-versa.

8.13 CALCULATING LINK-LEVEL ATTRIBUTES

Table 8.1 summarizes the reliability and maintainability ratings assigned to the modular equipment items that make up the radar and repeater sites and Control Center segments. From these values, the link-level parameters can be computed. For all of the equipment items except MUX Component A, calculating the failure rate contribution of the entity to the link aggregagate failure rate is a straightforward task. But, for MUX Component A, we must evaluate the beneficial effect conferred by incorporating one redundant unit in each entity. Six 6-channel units must be installed in each entity, in order to accommodate 36 transmission channels, but only five of these units must be operable at any given moment in order to satisfy the specified 30-channel criterion for determining that a message-interruption condition has not occurred. Therefore, we can treat the sixth unit of a MUX component A entity as redundant.

The other equipment items also appear to incorporate redundance in that they each possess a 36-channel capacity. Unfortunately, any failure within any equipment unit interrupts all of the channels carried by that unit. Since these other entities are made up of either 9-channel, 12-channel, or 36-channel unit modules, a unit failure within any one of these entities will reduce the capacity of that item to less than 30 channels—therefore, none of them can be treated as redundant.

Evidently, the manufacturer of the MUX chose to minimize the number of channels per Component A unit module because of the relatively high failure rate per channel of that item, 98.0 FPMH (failures per million hours), as compared with 13.3 FPMH for the Switching Panel, which is the equipment item with the next highest channel failure rate. The relatively low number of channels per Component A unit is fortuitous for our application because it enables us to offset the high channel failure rate with redundancy. The entry for MUX Component A in the last column of Table 8.1 shows the tremendous improvement in item reliability afforded by the redundant unit. The sum of the failure rates for the six units comprising the entity is 3528.0 FPMH, but the entity's effective failure rate is only 36.9 FPMH, a 99% reduction.

It must be emphasized that this enhancement of reliability can be reailized only if the redundancy is valid in terms of the function performed by the item in the specific application. In our example, the sixth unit would not be redundant if it were necessary for a technician to switch the extra unit into the circuitry in place of a failed unit, upon notification by an alarm signal, since the probability that the substitution would always be made within one minute is substantially less than 100%.

8.13.1 The Einhorn Equations

How was the effective failure rate given for MUX Component A in Table 8.1 calculated? In Chapter 7 we saw how failure rates are derived for combinations of two or more identical components operating in parallel such that the combination (which corresponds to an entity in our example) has not failed so long as any member of the combination remains operable. But we have a different situation here—only one of the six units is redundant—five of the six must be operable or the entity has failed. The method for calculating redundancy effects utilizes what have come to be called the *Einhorn equations*, after S. J. Einhorn, who formulated them. Tables 8.2 and 8.3 present general Einhorn equations for "equivalent up-time" and "equivalent down-time," respectively, and also tabulate specific equations for combinations of up to six units, where as few as one or as many as five units are redundant.

As we saw in Chapter 3, up-time and down-time are the dimensional terms used in calculating availability, and are analogous to MTBF and MTTR, respectively. However, when dealing with redundancy, a careful distinction must be made between these two pairs of corresponding terms. For MUX Component A, up-time applies to entity-level reliability, as enhanced by redundancy at the unit level. As the entries in Table 8.1 show, the effective MTBF of the entity and, consequently, of the link is made substantially greater by adding redundant units, even though the aggregate failure rate of the entity has thereby been increased. The up-time (Einhorn-equivalent MTBF) of the item is calculated by means of the equations in Table 8.2, and can be used for calculating reliabilities and availabilities at the link and higher assembly hierarchy levels.

In similar fashion, the Einhorn equations in Table 8.3 can be used for calculating the down-times utilized in determining availability at the entity and higher levels, but they cannot be used for calculating effective MTTR at any of these levels. Entity-level down-time includes only those periods during which the item is unavailable; MUX Component A is unavailable only when two or more units are inoperable concurrently. When only one unit is inoperable its down-time is not counted as entity-level down-time. But MTTR is defined in terms of the time required for restoring an inoperable hardware element to operability, and is the same regardless of whether the repair is performed on a redundant or non-redundant component of that element. Furthermore, since adding redundant units to an entity increases the entity's aggregate failure rate, the expected relative frequency of repair for that entity is increased accordingly.

The equations in Table 8.2 are used for determining the Einhorn up-time for a combination of "N" units, where at least "R" units must be operable at any given moment. We used them to calculate the Einhorn-equivalent failure rate for MUX Component A in Table 8.1.

8.13.1.1 Singling Out the Worst-Case Link

Because the link-level requirements (Paragraph 8.13.1) apply separately to every link, the simplest way to test the design against the requirements is to identify the worst-case link and calculate its RMA attributes. As it happens, there are very few variations in the links distinguishing one link from another. Most of the links interconnect two intermediate radar sites via a single repeater site. In those cases where one of the sites forming the link is a junction site or the Control Center, the link equipment complement is the

Table 8.2 Einhorn Mean Up-Time for R of N Identical Units with Repair

Einhorn Mean Up-Time *R* of *N*

N \ R	2	3	4
1	$\frac{2UD + U^2}{2D}$	$\frac{3UD^2 + 3U^2D + U^3}{3D^2}$	$\frac{4UD^3 + 6U^2D^2 + 4U^3D + U^4}{4D^3}$
2	X	$\frac{3UD + U^2}{6D}$	$\frac{6UD^2 + 4U^2D + U^3}{12D^2}$
3	X	X	$\frac{4UD + U^2}{12D}$
4	X	X	X
5	X	X	X

same as if both sites were intermediate radar sites, since the link must be capable of handling at least 30 channels in each direction accomodating the relay of messages between sites outside the link. With the exception of one dual-repeater link, every link includes one and only one repeater site. Therefore, the single-repeater configuration linking intermediate, junction, and/or Control Center sites can be designated as the "standard link configuration."

$$Us = \frac{\sum_{J=R}^{N} \binom{N}{j} U^j D^{N-j}}{N \binom{N-1}{R-1} U^{R-1} D^{N-R}}$$

5	6
$\frac{5UD^4 + 10U^2D^3 + 10U^3D^2 + 5U^4D + U^5}{5D^4}$	$\frac{6UD^5 + 15U^2D^4 + 10U^3D^3 + 15U^4D^2 + 6U^5D + U^6}{6D^5}$
$\frac{10UD^3 + 10U^2D^2 + 5U^3D + U^4}{20D^3}$	$\frac{15UD^4 + 20U^3D^3 + 15U^3D^2 + 6U^4D + U^5}{30D^4}$
$\frac{10\ UD^2 + 5U^2D + U^3}{30D^2}$	$\frac{20UD^3 + 15U^2D^2 + 6UD^3 + U^4}{60D^3}$
$\frac{5UD + U^2}{20D}$	$\frac{15UD^2 + 6U^2D + U^3}{60D^2}$
X	$\frac{6UD + U^2}{30D}$

Apart from the dual-repeater link, there is one other exception to the standard link configuration, and that is the link which interconnects a terminal radar site with another site (e.g., the link between the Control Center and Site G)—let's call this the "terminal link configuration." The terminal link includes fewer units per equipment item entity than the standard link, since such links are not required to relay messages between sites outside the link,

Table 8.3 Einhorn Mean Down-Time for R of N Identical Units with Repair

$$\text{Einhorn Mean Down-Time } R \text{ of } N \quad \text{Ds} = \frac{\sum_{j=0}^{R-1} \binom{N}{j} U^{j} D^{N-j}}{N \binom{N-1}{R-1} U^{R-1} D^{N-R}}$$

R \ N	2	3	4	5	6
1	$\frac{D}{2}$	$\frac{D}{3}$	$\frac{D}{4}$	$\frac{D}{5}$	$\frac{D}{6}$
2	X	$\frac{D^2 + 3UD}{6U}$	$\frac{D^2 + 4UD}{12U}$	$\frac{D^2 + 5UD}{20U}$	$\frac{D^2 + 6UD}{30U}$
3	X	X	$\frac{D^3 + 4UD^2 + 6U^2D}{12U^2}$	$\frac{D^3 + 5UD^2 + 10U^2D^2}{30U^2}$	$\frac{D^3 + 6UD^2 + 15U^2D}{60U^2}$
4	X	X	X	$\frac{D^4 + 5UD^3 + 10U^2D^2 + 10U^3D}{20U^3}$	$\frac{D^4 + 6UD^3 + 15U^2D^2 + 20U^3D}{60U^3}$
5	X	X	X	X	$\frac{D^5 + 6UD^4 + 15U^2D^3 + 20U^3D^2 + 15U^4D}{30U^4}$

but only to carry messages to and from the terminals in the terminal radar site. Because a terminal link includes fewer units, its aggregate failure rate is correspondingly lower and its MTBF and availability correspondingly greater. To identify the worst-case link configuration, we must return to the dual-repeater link, which interconnects the Control Center with Site U, since all other link configurations have been accounted for as either standard or terminal links.

8.13.1.2 The Negative Side of Redundancy

Table 8.4 presents reliability and maintainability values for calculating the MTBF, MTTR, and availability characteristics of the standard and worst-case link configurations. In Table 8.4, the paired entries for MUX Component A in the AGGREGATE FAILURE RATE Column bear the same relationship at the link level as the corresponding entries in the ENTITY FAILURE RATE Column of Table 8.1, both pairs demonstrating the 100-1 reduction in effective failure rate achieved by making the sixth unit redundant. But redundancy can have a negative side also, as illustrated by the following: the first of the two entries given for MUX Component A in the WEIGHTED REPAIR TIME Column is the product of the entity's MTTR and its link-level failure rate. Since each entity must be repaired every time that any one of its component units (including the redundant unit) fails, the contribution of MUX Component A to the maintenance support required by each link is a whopping 50,803.2 repair hours per million hours of link operation. If the redundant unit were removed from the MUX A entity, the repair time support burden would be reduced by a sixth.

The second entry for MUX Component A in the WEIGHTED REPAIR TIME Column of Table 8.4 (265.68 hours per million hours) is the product of the second entries in the AGGREGATE FAILURE RATE and ENTITY REPAIR TIME Columns for MUX Component A; this value is used only for calculating link and higher-level availabilities.

The entries in the AGGREGATE FAILURE RATE, ENTITY REPAIR TIME, and WEIGHTED REPAIR TIME Columns are totaled at the bottom of the table for the standard-case (single repeater) and worst-case link (dual repeater) configurations. The first and second entries in each of these column totals are for aggregate unit and Einhorn-equivalent values, respectively, of MUX Component A. Again it must be stressed that the second entry in the WEIGHTED REPAIR TIME Column of Table 8.4 is used only for the calculation of link and higher-level availabilities.

Table 8.4 Equipment Item Reliability and Maintainability Ratings for Calculating Link-Level Ratings

Equipment item	Quantity of entities per link[a]	Aggregate failure rate[b] (FPMH)	Entity repair time (hours)	Weighted repair time[c] (FPMH × hours)
Receiver	4	158.4	3.8	601.92
	6	237.6	3.8	902.88
Transmitter	4	172.8	3.8	656.64
	6	259.2	3.8	984.96
MUX Component A	4	14112.0	3.6	50803.20
	4	147.6	1.8	265.68
MUX Component B	4	129.6	2.1	272.16
Switching Panel	4	1915.2	3.8	7277.76
Amplifier, relay	2	446.4	4.8	2187.36
Amplifier, repeater	2	72.0	5.6	403.20
	4	144.0	5.6	806.40
Totals				
Standard link:				
Aggregate failure rate		17006.4[d]	3.66[e]	62202.24
Einhorn-equiv. fail. rate		3042.0	3.83	11664.72
Worst-case link:				
Aggregate failure rate		17244.0	3.67	63234.72
Einhorn-equiv. fail. rate		3279.6	3.87	12697.20

[a]For Receiver, Transmitter, and Repeater Amplifier, first (upper) entry in this column applies to the standard link configuration, second (lower) entry to the worst-case link configuration. (See Note b regarding dual entries for MUX Component A.)

[b]Aggregate contribution of equipment items entities to link failure rate for transmission in both directions. For MUX Component A, the first entry is aggregate failure rate of all units, second entry represents Einhorn equivalent failure rate.

[c]Weighted repair time for aggregate of equipment item entities included in the link configuration; calculated as the product of the entity repair time and link-level corresponding failure rate for the item, the latter weighting factor being proportional to relative frequency of repair. For MUX Component A only, second entry is Einhorn equivalent down-time, weighted by Einhorn-equivalent failure rate—used only for calculating link availability. (These values have not been converted to FPMH × one million hours, i.e., failures per hour, which would be the form compatible with other data, because these values are used only for calculating link-level MTTR, using a divisor also expressed in terms of FPMH.)

[d]First entry for worst-case link configuration is aggregate failure rate, including failure rate for four six-unit MUX Component A item entities (24 units total); second entry includes Einhorn equivalent failure rate for four MUX Component A item entries, instead of the aggregate failure rate for those entries.

[e]First entry for worst-case link configuration is MTTR, second entry is Einhorn-equivalent down-time; both entries obtained by dividing corresponding entry in WEIGHTED REPAIR TIME Column by corresponding entry in AGGREGATE FAILURE RATE Column. Second entry used only for calculating link availability.

The link-level MTTR and Einhorn-equivalent values in the ENTITY REPAIR TIME Column were derived by dividing the corresponding totals in the AGGREGATE FAILURE RATE Column into the matching totals in the WEIGHTED REPAIR TIME Column. Because of its extremely high aggregate failure rate, MUX Component A influences link-level MTTR more than all of the other equipment items combined. Since the MTTR for individual MUX Component A units is less than the specified maximum allowable link MTTR of 4.0 hours (Paragraph 8.3.1.3), it seems likely that this requirement will be satisfied, as indeed it is, the calculated values being 3.66 and 3.67 hours for the standard and worst-case link configurations, respectively. But the aggregate failure rate for the two link configurations translate into MTBF values of 328.7 and 304.9 hours, respectively, which come nowhere close to meeting the specified MTBF of 600 hours (Paragraph 8.3.1.1)—so we will have to add redundant units.

Before moving on to the task of selecting which equipment item is to be given redundant units, and how many, let's look at some other results of our link-level analysis. The total Einhorn-equivaleng entity repair times of 3.83 and 3.87 hours given in Table 8.4 are not much greater than the corresponding MTTR values of 3.66 and 3.67 hours, respectively. But if we take the reciprocals of the corresponding aggregate failure rates (upper totals in AGGREGATE FAILURE RATE Column), we arrive at values of 58.80 and 57.99 hours. What this means is that, if the sixth unit in each MUX Component A entity were not redundant, the link MTBF value would be less than one-fifth of the value predicted on the basis of redundancy.

The MTBF, MTTR, and WEIGHTED REPAIR TIME values calculated for the standard and worst-case link configurations serve to remind us once again of fundamental differences in the nature of these three units—the MTBF definitely tends to decrease as the number of series-connected components increases and the total maintenance manhours required to support the hardware tends to increase, but the MTTR may not be affected at all.

Even though we known that the link configuration will have to be improved in order to meet the link MTBF requirement, we should first check to see if we are also in trouble with the link availability requirement. As it turns out, we are—the availability of the worst-case link configurations is 98.75%—far short of the required 99.50%. Obviously, adding redundant units is our only option.

8.13.1.3 Selecting the Redundant Unit

In the majority of cases, the simplest and most cost-effective approach to enhancing reliability through redundancy is by an iterative procedure—first, identifying the lowest-level component with the highest failure rate; then, reinforcing that component with an identical redundant component. The procedure is repeated until the required reliability and availability values have been achieved. In the selection of the component to be reinforced, the relative quantities of the components per system must also be considered, because reinforcing a low-reliability component whose quantity per link is very small yields relatively little improvement in link reliability.

The logical candidate for reinforcement is the Switching Panel, since this equipment item is second only to the MUX Component A in magnitude of unit failure rate and contribution to link aggregate failure rate. Adding one unit to each Switching Panel, such that only four of the five Panel units need be operable at any given time for the Panel to accommodate 30 channels, reduces the Panel's Einhorn-equivalent failure rate at the link level from 1915.2 to 17.4 FPMH. (The Panel's link-level aggregate failure rate is increased by the addition of the fifth unit to 2394.0 FPMH but, since the Panel's repair time is 3.8 hours, the change in link-level MTTR is insignificant.) The MTBF for the worst-case link is increased, thereby, to 723.7 hours, which exceeds the specified minimum allowable value; worst-case availability rises to 99.46%. This does not quite meet the minimum specified value of 99.50% but is so close that the customer may, upon request by the system contractor, agree to grant a deviation to the specification requirement in order to avoid the costs of incorporating and maintaining any more redundant equipment units in the worst-case link. The contractor can give his argument still greater weight by pointing out that the worst-case configuration applies to only one link, and that the availability of each standard-configuration is 99.56%. (That of the terminal links is even better).

In our Communications Network example, we found it necessary to provide a redundant unit in the Switching Panels used in every link, in order to satisfy link MTBF and availability requirements. However, if the equipment reliability and repair time ratings had been better, it might have been necessary to reinforce only the worst-case link, thereby reducing greatly the cost of equipping the standard-configuration links. As was pointed out in Chapter 1, the optimal design for reliability is the balanced configuration, the configuration in which all circuits comprised of series-connected elements are equally reliable. The most cost-effective design is the one in which all circuits (or links) meet minimal requirements but none exceeds those requirements substantially.

(Incidentally, all of the calculations for this Communications Network example are shown in the appendix.)

8.13.2 The M/max/ct Equations

In the discussion of M/max/ct in Chapter 2, we saw how the value of this maintainability measure could be approximated for the stated percentile by plotting a cumulative percentage distribution curve of the repair times. The easiest way to derive the plotting data is to rearrange the aggregate failure rate data of Table 8.4 in order of increasing repair time, as shown in Table 8.5. The third column in Table 8.5, the Percentage of Total Failure Rate Column, lists for each equipment item the percentage contribution made by that item at the link level to the total link failure rate. The fourth column (Cumulative Failure Rate Percentage) provides the values for plotting the cumulative percentage distribution curve.

Tracing down the fourth column we find that the 95th percentile falls in the sector occupied by the Switching Panel. This would seem to indicate that the M/max/ct value would be less than 3.8 hours, since the repair time for the Switching Panel is 3.8 hours, and the repair times for all of the other items making up the zero-through-95th percentile group (i.e., the items listed above the Switching Panel in Table 8.5) are either 3.8 hours or less. But the indication is false, as we see when we actually draw the cumulative failure distribution curve for the standard link (Figure 8.4). From the data in the fourth column, we can see that the curve for the worst-case link will be almost identical.

8.13.2.1 A Shortcoming in Our Example

At this point it must be admitted that the data given for our Communications Network example is not representative of typical real-world situations. In the real world you can expect to find complex systems to be made up of a great many equipment items, and these items would exhibit a repair time distribution of successive tiny incremental changes such that plotting a histogram of this distribution would form a smooth, continuous curve. In contrast, our example provides only seven histogram blocks at the link level—to plot a frequency distribution curve or a cumulative frequency

Table 8.5 Link-Level Data Retabulated to Facilitate Approximating of M/max/ct for 95th Percentile

Equipment item	Entity repair time (hours)	Link-level item failure rate* (FPMH)	Percentage of total failure rate	Cumulative failure rate percentage
MUX Component B	2.1	129.6	0.74	0.74
	2.1	129.6	0.73	0.73
MUX Component A	3.6	14112.0	80.70	81.44
	3.6	14112.0	79.63	80.36
Receiver	3.8	158.4	0.91	82.35
	3.8	237.6	1.34	81.70
Transmitter	3.8	172.8	0.99	83.34
	3.8	259.2	1.46	83.16
Switching Panel	3.8	2394.0	13.69	97.03
(five-unit entity)	3.8	2394.0	13.51	96.67
Amplifier, relay	4.8	446.4	2.55	99.58
	4.8	446.4	2.52	99.19
Amplifier, repeater	5.6	72.0	0.41	99.99
	5.6	144.0	0.81	100.00
Totals				
Standard link:		17485.2	99.00	
Worst-case link:		17722.8	100.00	

*First (upper) entry for each equipment item is for standard link configuration; second (lower) entry is for worst-case link.

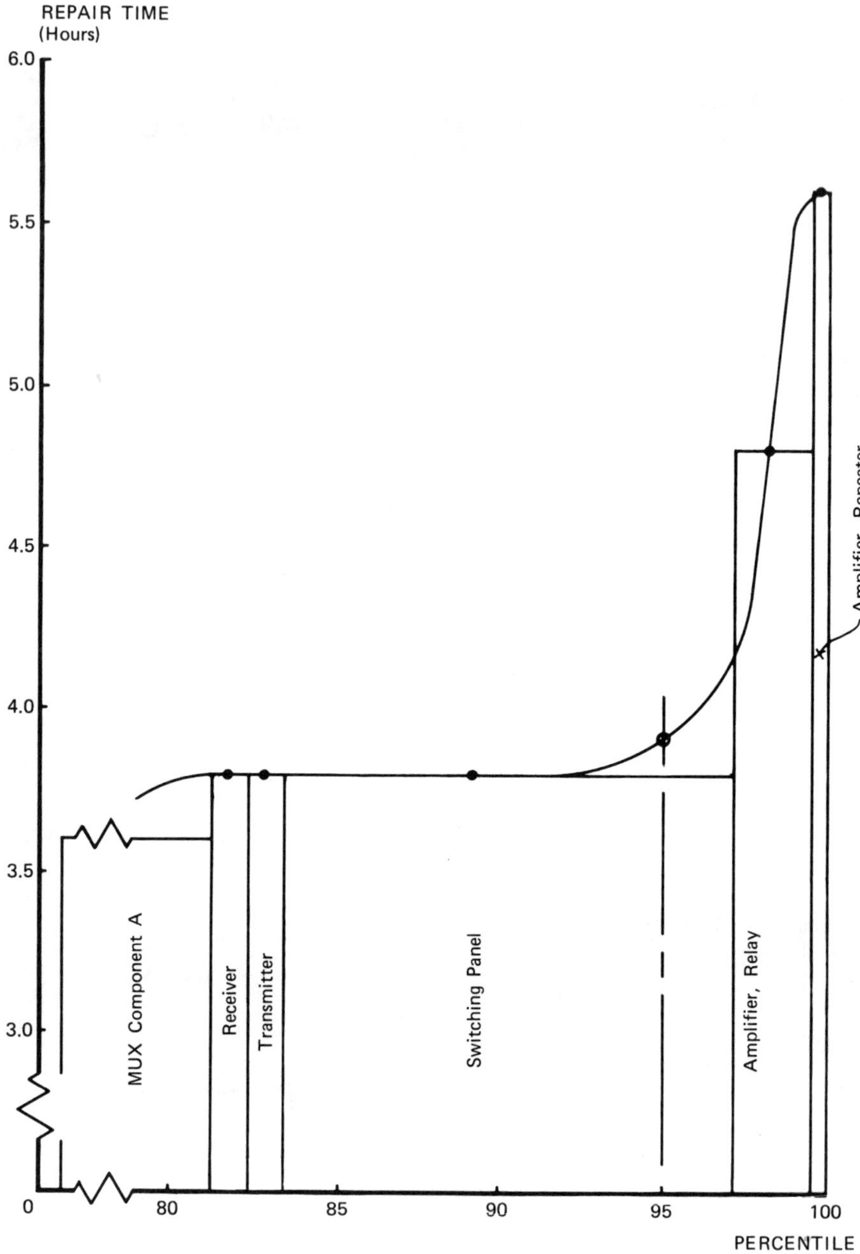

Figure 8.4 Cumulative failure rate percentage for standard link.

distribution curve, it is necessary to draw the curve through the midpoint of the very broad top of each block. Obviously, such curves cannot be drawn with any degree of precision.

But Figure 8.4 does show us why the M/max/ct value can be expected to be greater than 3.8 hours. As the cumulative distribution curve moves from left to right in traversing the Switching Panel block, it is below the 3.8-hour level until it reaches the midpoint of the block, but rises above the 3.8-hour level thereafter. Since the midpoint is just slightly beyond the 90th percentile, the 95th percentile falls in the region where the curve is above the 3.8-hour level. But we know that the plotting of the curve is not accurate, and is particularly suspect in the vicinity of the 95th percentile because the curve rises very rapidly in this sector.

To put our minds at ease, it will be necessary to use the messy equations given in Chapter 2 to calculate M/max/ct. Details of the calculations are set forth in the appendix. The equations yield M/max/ct values of 3.88 and 3.89 hours for the standard and worst-case links, respectively—not much greater than the corresponding MTTR values, and well within the specified maximum allowable M/max/ct of 6.0 hours. Again, these results are not representative of typical real-world situations, but stem from the data chosen for the Communications Network example.

8.13.2.2 Reducing Excessive MTTR Values

If either the MTTR or the M/max/ct had exceeded their maximum allowable values, it would be necessary to re-evaluate the design in order to find the most cost-effective method of improving its maintainability. By examining the list of time increments making up the total time for each repair task, the lengthiest increments can be identified as candidates for maintainability improvement. The candidates selected should then be ranked in order of ease of improvement, where ease of improvement is inversely proportional to the extent of hardware redesign required. Generally, the easiest task increment times to reduce are those involving lengthy fault isolation tasks.

In some cases, fault isolation can be speeded up by rewriting inefficient procedures, in other cases, incorporation of built-in test equipment may be needed. Overlong access or removal/replacement times can sometimes be shortened by substituting one kind of fastener with another, threaded hose couplings with quick-disconnect couplings, or soldered-connection wiring ("hard wiring") with plug-in connectors. It is almost never feasible to attempt fundamental design changes; if the manufacturer of the item has not designed it in accordance with maintainability design principles (standardization, modularization, etc.), the only alternative is to find an alternate equipment source. But, if system-level maintainability principles had been followed from the outset, equipment items with poor maintainability characteristics would never have been included.

8.14 CALCULATING SYSTEM-LEVEL ATTRIBUTES

The system-level requirements set forth under Paragraph 8.3.2 address communications between the Control Center and individual radar sites. In the course of determining whether or not the design satisfies these requirements, we will again simplify matters by calculating system-level parameters for the worst-case situations rather than trying to compute the MTBF and availability for every possible pairing of Control Center and radar site.

8.14.1 From Figure 8.1 we can see that, when the Control Center is paired with a radar site that is one of a series-linked set of sites (e.g., the communications line terminating in Site J, which comprises Leg VII in Figure 8.3), the further the site is from the Control Center (i.e., the greater the number of intervening sites), the worse the reliability/availability case. Thus, if distance between Control Center and site were the only consideration, the worst case would be the pairing of the Control Center with the radar site furthest from it. But distance is not the only factor, as we will see a little further on.

Figure 8.1 shows the Control Center as being interconnected with the radar sites via circuits composed of various quantities of only a few types of sites. Since sites of the same type are identical, with regard to reliability and maintainability measurements, the calculations necessary for determining whether or not the design satisfies system-level requirements can be facilitated by tabulating the site type data. Four types of sites make up the Control-Center-to-radar-site circuits:

1. Control Center terminal segment or radar terminal site; see Table 8.6(a)
2. Intermediate radar site or junction radar (relay segment) site; see Table 8.6(b)
3. Repeater site; see Table 8.6(c)
4. Junction radar (relay segment); see Table 8.6(d)

Table 8.6a Equipment Item Reliability and Maintainability Ratings for Calculating System-Level Ratings

	SITE: Control Center (segment) or Terminal Radar				
Item	Quantity of units per site (Note 1)	Aggregate failure rate (FPMH)	Einhorn-Equivalent failure rate (FPMH)	Entity repair time (Hours)	Weighted repair repair time (FPMH × hours)
Receiver	1	13.2	13.2	3.8	50.16
Transmitter	1	14.4	14.4	3.8	54.72
MUX Component A	2	1176.0	73.8	3.6	4233.60
	2	1176.0	73.8	1.8	132.84
MUX Component B	2	64.8	64.8	2.1	136.08
Switching Panel	2	239.4	4.4	3.8	909.72
	2	239.4	4.4	1.9	8.36
Totals:			(Note 2)		
Aggregate:		1507.8			5384.28
Einhorn-equivalent:			170.6		382.16

Notes:

1. Quantity of Units required per site to provide at least six channels in each of two directions for message transmission. (Equipment installed in unit form, rather than as multiple-unit entities—every equipment item unit can accommodate at least six channels.)
2. Column total is sum of Einhorn-equivalent failure rate for MUX component A and Switching Panel, where applicable, and aggregate failure rates for all items where not applicable.

Table 8.6b Equipment Item Reliability and Maintainability Ratings for Calculating System-Level Ratings

	Site: Intermediate Radar or Junction Radar (relay)				
Item	Quantity of units per site (Note 1)	Aggregate failure rate (FPMH)	Einhorn-equivalent failure rate (FPMN)	Entity repair time (hours)	Weighted repair time (FPMH × hours)
Receiver	6	79.2	79.2	3.8	300.96
Transmitter	6	86.4	86.4	3.8	328.32
MUX Component A	24	14112.0	147.6	3.6	50803.20
	24	14112.0	147.6	1.8	265.68
MUX Component B	4	129.6	129.6	2.1	272.16
Switching Panel	16	2394.0	17.4	3.8	9097.20
	16	2394.0	17.4	1.9	33.06
Amplifier, relay	6	446.4	446.4	4.8	2142.72
Totals:			(Note 2)		
Aggregate		17247.6			62944.56
Einhorn-equivalent			906.6		3342.90

Notes:

1. Quantity of units required per site to provide at least 30 channels in each of two directions for message transmission.
2. Column total is sum of Einhorn-equivalent failure rate for MUX Component A and Switching Panel, where applicable, and aggregate failure rates for all items where not applicable.

Table 8.6c Equipment Item Reliability and Maintainability ratings for Calculating System-Level Ratings

	Site: Repeater				
Item	Quantity of units per site (Note 1)	Aggregate failure rate (FPMH)	Einhorn-Equivalent failure rate (FPMN)	Entity repair time (hours)	Weighted repair time (FPMH × hours)
Receiver	6	79.2	79.2	3.8	300.96
Transmitter	6	86.4	86.4	3.8	328.30
Amplifier, repeater	6	72.0	72.0	5.6	403.20
Totals:			(Note 2)		
Aggregate:		237.6			
Einhorn-equivalent:			237.6		1032.48

Notes:

1. Quantity of units required per site to provide at least 30 channels in each of two directions for message transmission.
2. Column total is sum of Einhorn-equivalent failure rate for MUX Component A and Switching Panel, where applicable, and aggregate failure rates for all items where not applicable.

Table 8.6d Equipment Item Reliability and Maintainability Ratings for Calculating System-Level Ratings

	SITE: Junction Radar (segment)				
Item	Quantity of units per site (Note 1)	Aggregate failure rate (FPMH)	Einhorn-Equivalent failure rate (FPMH)	Entity repair time (hours)	Weighted repair time (FPMH × hours)
Receiver	3	39.6	39.6	3.8	150.48
Transmitter	3	43.2	43.2	3.8	164.16
MUX Component A	6	3528.0	36.9	3.6	12700.80
	6	3528.0	36.9	1.8	66.42
MUX Component B	1	32.4	32.4	2.1	68.04
Switching Panel	4	598.5	4.4	3.8	2274.30
	4	598.5	4.4	1.9	8.36
Amplifier, relay	3	223.2	223.2	4.8	1071.36
Total			(Note 2)		
Aggregate		4464.9			16429.14
Einhorn-equivalent			379.7		1528.82

Notes:

1. Quantity of units required per site to provide at least 30 channels in each of two directions for message transmission.
2. Column total is sum of Einhorn-equivalent failure rate for MUX Component A and Switching Panel, where applicable, and aggregate failure rates for all items where not applicable.

This grouping of types requires a bit of explanation. A Control Center terminal segment and a radar terminal site are comprised of identical equipment complements that serve only the on-site terminals at which messages originate or terminate. Six channels in each direction suffice for this type of site, since it does not relay messages between other sites.

That portion of a junction radar site which serves to tie two legs into a single path is identical to an intermediate radar site, each relaying messages between other sites, as well as incorporating on-site terminals—30 channels are needed in either direction.

8.14.2 Dealing with Redundant Communication Paths

The repeater site type need not be described again, but the junction radar relay segment requires a bit of explanation. Analysis of pairings between the Control Center and most radar sites is complicated by the fact that messages between the pair can follow more than one path. For example, Site A is interconnected with the Control Center via three different paths, each of which traverses two or more legs joined by junction radar sites (see Figure 8.3). Calculation of MTBF and availability for each pairing is performed as follows:

1. Failure rates and repair times are compiled for each leg—Table 8.7 lists the quantities of each site type making up each leg, and Table 8.8 lists the failure rates and repair times for each leg. Junction sites are excluded from the tabulated leg data to avoid computing confusion arising in multi-path pairing situations.

Although calculation of MTBF and availability values requires only the Einhorn-equivalent failure rates and weighted Einhorn-equivalent repair times, Table 8.8 also presents aggregate failure rates and weighted MTTRs, to permit estimation of system-level MTTRs and total maintenance support manpower requirements for logistic planning purposes.

2. For any given Control-Center-to-radar-site pairing interconnected by only one path (e.g., the pairing with Radar Site J), the Einhorn-equivalent failure rate and weighted repair time are the corresponding values for Leg IX—the path forms a series-connected circuit).

3. For a pairing interconnected by more than one path, the paths can form a series-parallel circuit. The Control Center pairing with Terminal Site A is just such a circuit, and it is dealt with by successively combining parallel paths until an equivalent series-only

Table 8.7 Site Equipment Complement for Communications Network Legs*

Leg	Terminal radar or control center	Intermediate radar	Repeater
I	1	0	1
II	1	4	5
III	0	2	3
IV	1	3	4
V	0	1	2
VI	1	1	2
VII	1	0	1
VIII	2	0	1
IX	2	2	3
X	1	1	2
XI	1	2	4
XII	1	0	1
XIII	1	0	1

*For any leg terminating in a junction radar site, the junction is not counted as part of the leg site complement.

circuit is achieved. For each successive combination, the Einhorn-equivalent failure rate and weighted repair time are calculated until the values for the pairing have been reached. The sequence in which the combinations are to be resolved is in order of relative subordination, the lowest level combinations to be taken earliest.

(a) The first combination to be resolved for the Terminal Site A pairing consists of the two paths interconnecting the Control Center with Junction Site M. Leg IV constitutes one of these paths; Legs V and VI, plus Junction Site R (relay) and Junction Site M (segment) to other.

It is important to understand how the junction sites are handled in this combination—Junction R (relay) is treated as if it were an intermediate radar site linking the sites in Legs V and VI with which it mates, whereas Junction M (segment) links its mating site in Leg V to the rest of Junction M.

(b) The second combination consists of the two paths connecting the Control Center to Junction Site B. Leg II plus Junction

Table 8.8 Reliability and Maintainability Ratings for Communications Network Legs*

Leg	Aggregate failure rate (FPMH)	Einhorn-equivalent failure rate (FPMH)	Weighted repair time (FPMH × hours)	Weighted Einhorn-equivalent repair time (FPMH × hours)
I	1745.4	408.2	6416.76	1414.64
II	71686.2	4985.0	262324.92	18916.16
III	35208.0	2526.0	128986.56	9783.24
IV	54201.0	3840.8	198347.88	14540.78
V	17722.8	1381.8	65009.52	5407.86
VI	19230.6	1552.4	70393.80	5790.02
VII	1745.4	408.2	6416.76	1414.64
VIII	3253.2	578.8	11801.04	1796.80
IX	38223.6	2867.2	139755.12	10547.56
X	19230.6	1552.4	70393.80	5790.02
XI	36953.4	2934.2	135403.32	11197.88
XII	1745.4	408.2	6416.76	1414.64
XIII	1745.4	408.2	6416.76	1414.64

*Leg parameters do not include terminations in junctions.

Site B (segment) constitutes one path; Leg III plus Junction Site M (relay) plus the first combination the other path.

(c) The interconnection pairing has been reduced to a series-only circuit consisting of the second combination plus Junction Site B (relay) plus Leg I.

In evaluating the Communications Network for worst-case pairings, we can utilize the following strategy: First, only those pairings are considered that interconnect the Control Center with a terminal radar site, since each terminal site is further away than any other site along the interconnecting path. Second, the candidate pairings are grouped according to how many paths are included in the interconnection. Third, within each path group, a screening can usually made by evaluating the configurations of its members (Figure 8.1) to identify the pairing with the largest site population. (This evaluation can also make use of the leg data in Table 8.8.)

For our example, this strategy provides us with three candidates, Terminal Site J (single path), Terminal Site Y (two paths), and Terminal Site A (three paths). Their respective MTBF's prove to be 348.8, 740.4, and 237.4 hours—Terminal Site A is our worst case, but it does substantially exceed the 200 hour minimum allowable value specified in Paragraph 8.3.2.1. The availability of the Terminal Site A pairing is 99.11%, which exceeds the 97.50 minimum allowable value in Paragraph 8.3.2.2.

8.15 SOME CONCLUSIONS

Our analysis has enabled us to accomplish two things: One—it disclosed a reliability deficiency at the link level, and indicated the optimal method for correcting this deficiency. Two—it provided quantitative evidence that the design, as modified, would satisfy all specified RMA requirements. This assurance can be very important to your own company, as well as to your customer, when a large-scale product development project is still in the planning stage.

In the course of performing the analysis we made certain judgements as to the customer's reasoning in formulating his RMA specifications, and the influence these judgements had on the manner in which we strove to satisfy the specification requirements. At both the link and system levels, the requirements address the worst-case situation. Since the worst-case two-repeater link is unique, all of the other links can be expected to demonstrate substantially better MTBF and availability performance. Similarly, communication between the Control Center and most radar sites should be more reliable than the pairing of the Center with Terminal Site A. Thus, the design possesses a safety margin against unforseen problems.

APPENDIX

As promised in the main text of Chapter 8, mathematical calculations performed in the course of determining link- and system-level RMA ratings are set forth herein. The first calculation is that of the effective failure rate for the six-unit MUX Component A, as it appears in Table 8.1 (see Section 8.13 in the text). Since only five of the six units must be operational in order to meet the 30-channel minimum transmission requirement, the appropriate equation in Table 8.2 for calculating Einhorn mean up-time of the six-unit entity is the equation for a "R" value of 5 and an "N" value of 6. This equation is:

$$E_u = \frac{6UD + U}{30D}$$

where

E_u is Einhorn mean up-time for MUX Component A entity

U is up-time (MTBF) of single unit, and equals $1/\lambda_u$ (reciprocal of unit failure rate), which is 1700.7 hours.

D is down-time (MTTR) of single unit, which is 3.6 hrs.

$$E_u = \frac{6 \times 1700.7 \times 3.6 + (1700.7)}{30 \times 3.6} = 27120.8 \text{ hrs.}$$

Entity failure rate = 1/27120.8 = 36.9 FPMH

The entity repair time of 1.8 hours shown in Table 8.4 for the MUX Component A is calculated in the same way, using the Einhorn mean down-time equation shown in Table 8.3 for an "R" value of 5 and an "N" value of 6. The same general procedure is followed for calculating the Einhorn-equivalent up and down-times for the redundant unit selected as described in Section 8.13.1.3.

Incidently, the derivation of the fundamental Einhorn equation was described in a paper by S. J. Einhorn, titled Reliability Prediction for Repairable Redundant Systems, which appeared in the

February 1963 Proceedings of the IEEE (Pages 312 to 317). Unfortunately, I have been unable to locate the original source of the specific set of equations derived from the fundamental equation that are presented in Tables 8.2 and 8.3.

The next set of calculations are for the link-level ratings derived from the data in Table 8.4, and discussed in Section 8.13.1.2; we'll look at the calculations for the worst-case link (one intermediate radar site plus two repeaters):

$$U_l = 10^6/\lambda_l$$

where

U_l is Link MTBF (up-time)

λ_l is Link Einhorn-equivalent failure rate (FPMH)

$$U_l = 10^6/3279.6 = 304.9 \text{ hrs.}$$

$$D_l = M_l/U_l$$

where

D_l is Link Einhorn-equivalent down-time (for availability calculation only)

M_l is Link Einhorn-equivalent repair time

$$D_l = 12697.2/3279.6 = 3.87 \text{ hrs.}$$

$$A_l = \frac{U_l}{U_l + D_l}$$

where

A_l is Link availability

$$A_l = \frac{304.9}{304.9 + 3.87} = 0.98747$$

$$MTTR_L = W_L/\lambda_L$$

where

$MTTR_L$ is link aggregate MTTR

W_L is link aggregate weighted repair time

λ_L is link aggregate failure rate

The standard link ratings are calculated in the same way. Using the M/max/ct equations of Section 2.4 to obtain the maximum corrective maintenance time for the standard or worst-case link (see Section 8.13.2.1) entails a great deal of calculation, as we are about to see.

$$M/max/ct = \text{antilog}_e\ (A + B)$$

$$A = \frac{\sum_{i=1}^{N_c} \ln X_{ci}}{N_c}$$

$$B = \psi \sqrt{\frac{\sum_{i=1}^{N_c} (\ln X_{ci})^2 - \left[\left(\sum_{i=1}^{N_c} \ln X_{ci}\right)^2 / N_c\right]}{N_c - 1}}$$

Where: N_c is the total number of repair tasks performed, X_{ci} is the time expended in performing the 'i'th task, and ψ is a coefficient whose value depends on the percentile chosen as the maximum repair time boundary (for 95th percentile, $\psi = 1.645$).

The most convenient way of organizing the calculations required is to tabulate the subordinate data elements as they are computed—this is done below:

Table of Computed Data Elements For Standard and Worst-Case Link Incorporating Five-Unit Switching Panel

Item	X_{ci}	N_c	$\ln X_{ci}$	$\sum_{i=1}^{N_c} \ln X_{ci}$	$(\ln X_{ci})^2$	$\frac{(\ln X_{ci})^2}{N_c}$
Receiver	3.8	158.4	1.335	211.46	1.782	282.27
	3.8	237.6	1.335	317.20	1.782	423.40
Transmitter	3.8	172.8	1.335	230.69	1.782	307.93
	3.8	259.2	1.335	346.03	1.782	461.89
MUX-A	3.6	14112.0	1.281	18077.47	1.641	23157.79
MUX-B	2.1	129.6	0.742	96.16	0.550	71.28
Switching	3.8	2394.0	1.335	3195.99	1.782	4266.11

Table (Continued)

Item	X_{ci}	N_c	$\ln X_{ci}$	$\sum_{i=1}^{N_c} \ln X_{ci}$	$(\ln X_{ci})^2$	$\frac{(\ln X_{ci})^2}{N_c}$
Panel						
Amplifier, relay	4.8	446.4	1.509	700.40	2.462	1099.04
Amplifier, repeater	5.6	72.0	1.723	124.06	2.969	213.77
	5.6	144.0	1.723	248.11	2.969	427.54
Totals						
Standard link		17485.2		22013.23		29398.19
Worst-case link		17722.8		22358.36		29907.05

Note: The entries in this table are made as follows:

a. Each entry in the first column of values (X_{ci}) is the time required for performing the corrective maintenance task for the corresponding equipment item; the MTTR value is used.

b. Each entry in the second column of values (N_c) is the relative frequency with which that task is expected to be performed; the failure rate (λ) is used. Two lines of values are given for the Receiver, Transmitter, and Repeater Amplifier; the values in the upper line are for the standard link—those in the lower line are for the worst-case link.

c. Each entry in the third column of values ($\ln X_{ci}$) is the natural log of the corresponding value in the first value column.

d. Each entry in the fourth column of values ($\Sigma \ln X_{ci}$) is the summation of all corrective maintenance task times, from the first task time ($i =$ 1 to the last task time $i = N$) for the corresponding equipment item. Since all of the task times for a given item are assumed to be the same, in this example, the entry is derived as the product of the corresponding entries in the second and third value columns.

e. Each entry in the fifth column of values $(\ln X_{ci})^2$ is the square of the corresponding entry in the third value column.

f. Each entry in the sixth column of values ($(\ln X_{ci})^2/N_C$) is the result of dividing the corresponding entry in the fifth value column by the corresponding entry in the second value column.

g. The entries in the second, fourth, and sixth value columns are summed at the bottom of the table to yield the entries needed in the M/max/ct equation set. (The entries in the first, third, and fifth value columns are included in the table only for computing the entries in the second, fourth, and sixth value columns.) Let's calculate the M/max/ct for the worst-case link:

$$A = 22.358.4/17722.8 = 1.26159$$

$$B = 1.645 \frac{29907.0 - (22358.4)/17722.8}{17722.8 - 1}$$

$$B = 0.09596$$

$$M/max/ct = \text{antilog}_e\ 1.3575 = 3.887 \text{ hrs.}$$

Index